Biosensor Design
and Application

ACS SYMPOSIUM SERIES **511**

Biosensor Design and Application

Paul R. Mathewson, EDITOR
Nabisco Foods Group

John W. Finley, EDITOR
Nabisco Foods Group

Developed from symposia sponsored
by the Biotechnology Secretariat
(Divisions of Agricultural and Food Chemistry,
Agrochemicals, Biochemical Technology,
and Small Chemical Businesses)
at the 201st National Meeting
of the American Chemical Society,
Atlanta, Georgia,
April 14–19, 1991

American Chemical Society, Washington, DC 1992

Library of Congress Cataloging-in-Publication Data

Biosensor design and application /   Paul R. Mathewson, editor, John W. Finley, editor.

   p.   cm.—(ACS symposium series, ISSN 0097–6156; 511).

"Developed from symposia sponsored by the Biotechnology Secretariat . . . at the 201st national meeting of the American Chemical Society, Atlanta, Georgia, April 14–19, 1991."

Includes bibliographical references and index.

ISBN 0–8412–2494–3

1. Biosensors—Congresses.

I. Mathewson, Paul R. II. Finley, John W., 1942–   . III. Series.

R857.B54B536   1992   92–31003
681′.2—dc20
                              CIP

Foreword

THE ACS SYMPOSIUM SERIES was first published in 1974 to provide a mechanism for publishing symposia quickly in book form. The purpose of this series is to publish comprehensive books developed from symposia, which are usually "snapshots in time" of the current research being done on a topic, plus some review material on the topic. For this reason, it is necessary that the papers be published as quickly as possible.

Before a symposium-based book is put under contract, the proposed table of contents is reviewed for appropriateness to the topic and for comprehensiveness of the collection. Some papers are excluded at this point, and others are added to round out the scope of the volume. In addition, a draft of each paper is peer-reviewed prior to final acceptance or rejection. This anonymous review process is supervised by the organizer(s) of the symposium, who become the editor(s) of the book. The authors then revise their papers according the the recommendations of both the reviewers and the editors, prepare camera-ready copy, and submit the final papers to the editors, who check that all necessary revisions have been made.

As a rule, only original research papers and original review papers are included in the volumes. Verbatim reproductions of previously published papers are not accepted.

M. Joan Comstock
Series Editor

Contents

Preface

A VIRTUAL EXPLOSION OF PAPERS IN THE BIOSENSOR FIELD reflects the extraordinary diversity of biosensor design and application. In fact, if we examine the number of biosensor citations during the past decade, we see a nearly exponential increase in the number of papers published. Undoubtedly, many factors—including the development of new types of chemical materials, miniaturization technology as exemplified by the computer chip industry, and the need for more directed and specific measurements in complex matrices—contributed to the increased interest in this form of miniaturized instrumentation.

The biosensor field is growing so rapidly and has become so diverse that it is impossible to comprehensively cover the entire field in one publication. Biosensors were a major theme at the meeting on which this book is based. Topics ranged from the basic chemistry through applications and commercial opportunities. A sister publication to this volume (*Biosensors & Chemical Sensors: Optimizing Performance Through Polymeric Materials,* edited by Peter G. Edleman and Joseph Wang, ACS Symposium Series 487, 1992) covers the basic chemistry and polymer science of biosensors. The purpose of this volume is to present a contemporary discussion of currently available and potential applications of biosensors.

The chapters in this book demonstrate the diverse approaches to the basic problem of microsizing analytical instrumentation. The methods run a gamut from use of intact crustacea antenna (perhaps the true BIOsensor) to physical spectroscopic means of coaxing both biological and physical substances into service as analytical tools. These methods include various forms of redox reactions (both enzymatic and nonenzymatic), those based on immunochemistry (utilizing antibody-mediated reactions), and several examples of optical- and spectrophotometric-based methods. In addition, several chapters deal with commercial and special situation uses.

Chapters 1–5 deal with a number of methods that take advantage of various types of oxidation and reduction reactions as the basis for analysis. These chapters demonstrate several strategies using the specificity of enzymic reactions to provide selective, sensitive analyses. A different strategy is demonstrated in the use of an organic metalloporphyrin for reduction of organohalides. These methods have found application in the agriculture and food area as well as in medicinal and clinical work.

Chapters 6–9 cover methods based on the use of antibodies. These methods incorporate redox or other common biological reactions but they introduce a high degree of selectivity because they include antibody binding. The means of detection can be spectroscopic, fluorometric, amperometric, or other methods. Thus, specific analytes can be assayed within a complex matrix.

Then spectrophotometric methods of analysis are covered in Chapters 10–13 presenting some rather unique and interesting optical-sensing methods, particularly the relatively new spectroscopic techniques of surface plasmon resonance and near infrared. These techniques use biological molecules to generate the selectivity and optical methods to monitor the various reactions.

The last two chapters describe some requirements for using biosensors in experiments conducted in space as well as an interesting approach to the development of more efficient procedures for commercialization of biosensor technologies.

Several chapters cover material that crosses over the arbitrary boundaries that we set up for each topic. A method using fluorescence as the detection system may be based on an antibody-mediated reaction, or oxidation–reduction reactions may be monitored spectrophotometrically. The variety of techniques evidenced in this volume may prompt the question of whether all techniques can be properly referred to as biosensors. Intuitively, the term "biosensor" implies that the means of sensing is biological, that is, derived from a natural process rather than from a physical–mechanical process. Taking this further, we might say that the sensing element is, in some way, a biomolecule rather than a mechanical instrument in which, for example, light, separated into its component wavelengths by means of a grating, impinges on a molecule (biological or synthetic) and is detected by a photooptical device. However, as you will read, the sensor component is not necessarily what is biological. The sensor may be optical, electrical, or mechanical, responding to electron transfer in a biological medium. One of the authors in this volume describes a biosensor as a "device incorporating a biological sensing element with a physical transducer" (Rodgers et al.), while another describes it as having two elements; "the molecular recognition element is the biological component of the biosensor that provides the device with a degree of chemical selectivity. The transducer, on the other hand, is the non-biological portion . . ." (Buch). It would appear from the chapters included in this volume that while these definitions fit some of the "microinstruments" discussed, they are perhaps a bit too restrictive. After digesting the melange of ideas contained in this work, we suggest the following more encompassing, yet usefully restrictive, definition: "A biosensor is an analytical device containing a biologically active component that serves to mediate a quantifiable chemical reaction or interaction."

Clearly, biosensors encompass an extremely diverse number of methods and applications. In the production environment, biosensors can be used for on-line measurement of critical intermediates or products; in the medical field, they can be used for rapid and economical determination and monitoring of metabolites, drugs, or hormones; and biosensors can be used as microsensing and control devices in the service of environmental, agricultural, and food-processing applications. We noted the exponential growth in interest in biosensors and believe that interest will continue to increase in the future. Biosensors offer a tremendous potential for simple, rapid, and accurate measurement of a virtually limitless number of analytes in an equally diverse matrix of applications. We hope that this book will serve as a useful record of current applications and a catalyst for future developments.

PAUL R. MATHEWSON
Nabisco Foods Group
200 Deforest Avenue
East Hanover, NJ 07936

JOHN W. FINLEY
Nabisco Foods Group
200 Deforest Avenue
East Hanover, NJ 07936

April 3, 1992

Chapter 1

Catalysis and Long-Range Electron Transfer by Quinoproteins

Victor L. Davidson

Department of Biochemistry, University of Mississippi Medical Center, Jackson, MS 39216-4505

Quinoproteins are a newly characterized class of enzymes which use novel quinones as redox cofactors. Quinoprotein dehydrogenases recognize a wide range of alcohols, amines, and sugars as substrates. They do not require other soluble cofactors or react directly with oxygen, and normally donate electrons to other protein-bound redox centers. Long range intermolecular electron transfer occurs between the redox centers of a quinoprotein, methylamine dehydrogenase, and a copper protein, amicyanin. Kinetic, biochemical, and biophysical techniques have been used to describe factors which are relevant to this process. The data obtained from studies of these proteins and the complex which they form are providing a picture of how nature stabilizes interactions at protein-macromolecule interfaces, and accomplishes long range electron transfer through proteins. Understanding precisely how this process occurs is critical to the development of direct amperometric biosensors, and will hopefully aid in the logical design of enzyme-based analytical devices.

A promising potential exists for the use of enzyme-based biosensors for the selective and sensitive detection of a wide variety of compounds. Such devices would require relatively stable enzymes which could efficiently be coupled to an electronic sensor. It would be most desirable in such a device to employ an enzyme which could transfer electrons directly to an electrode. This would require that it be a redox enzyme, one which undergoes oxidation and reduction as a part of its catalytic cycle, and which could ideally mediate the transfer of electrons from the substrate being measured directly to an electrode. The development of such direct amperometric biosensors has been

0097–6156/92/0511–0001$06.00/0

difficult, however, in large part because of properties of the redox enzymes which have been routinely used for this purpose.

Redox enzymes have traditionally been divided into two major classes, pyridine nucleotide-dependent dehydrogenases and oxidases. Neither type of enzyme is particularly well suited for use in a direct amperometric biosensor. The former requires a soluble coenzyme, either nicotinamide-adenine dinucleotide or nicotinamide-adenine dinucleotide phosphate, as an electron acceptor. These coenzymes are expensive, unstable and able to freely diffuse through the medium, and they must be regenerated to participate in further catalytic events. Oxidases pose the disadvantage of using oxygen as an electron acceptor and cosubstrate. Furthermore, the vast majority of these two categories of redox enzymes are fairly labile under conditions in which biosensors are likely to be used. For use in a biosensor, an ideal alternative enzyme would be relatively stable, possess a tightly bound redox center, use a modified electrode as a reoxidant, exhibit high turnover numbers, and require no additional soluble cofactors. Several members of a newly recognized class of enzymes, quinoproteins (1-3), appear to meet these criteria.

Properties of Quinoproteins

Quinoproteins utilize novel quinone species as cofactors and prosthetic groups. Three such quinones have been characterized thus far (Figure 1). Most bacterial quinoproteins possess pyrroloquinoline quinone [PQQ], the structure of which was characterized from the isolated cofactor of methanol dehydrogenase (4). PQQ is non- covalently but very tightly bound, and in all cases requires denaturation of the enzyme for its removal. A bacterial enzyme, methylamine dehydrogenase, possesses a covalently-bound tryptophan tryptophylquinone [TTQ] prosthetic group which is derived from a posttranslational modification of two gene encoded tryptophan residues (5,6). A third quinone redox prosthetic group, topaquinone, has been identified in mammalian plasma amine oxidase (7). It is also covalently-bound and derived from a posttranslational modification of a gene encoded tyrosine residue.

Although some quinoproteins are oxidases, most of the bacterial quinoproteins which have been characterized thus far are dehydrogenases (8-18), and catalyze the oxidation of a wide range of substrates (Table 1). The sensitive detection of many of these compounds which serve as substrates for these enzymes would be applicable to a wide range of disciplines including medicine, fermentation, food processing and environmental studies. Most of the enzymes which have been characterized are relatively easy to purify, relatively stable, and can be produced in large quantities by the host organism when grown under appropriate conditions. Furthermore, quinoprotein dehydrogenases do not appear to require additional soluble cofactors. Depending upon the enzyme, the natural electron acceptor is either a cytochrome, a copper protein, or a membrane-bound ubiquinone (19). Thus, these enzymes normally donate electrons to a redox center within a complex macromolecular matrix and efficient long range electron transfer, through the protein, is required for their physiological functions. The best characterized quinoprotein with respect to its redox and electron transfer properties is methylamine dehydrogenase from *Paracoccus denitrificans*. This paper will

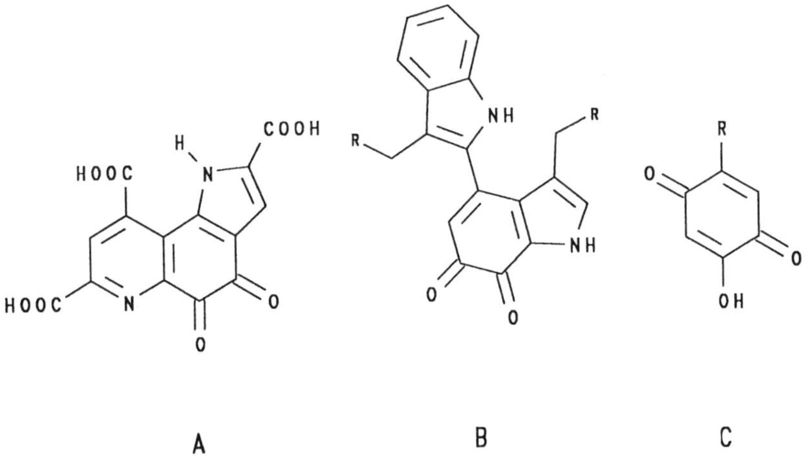

Figure 1. The structures of quinone prosthetic groups. A. Pyrroloquinoline quinone [PQQ]. B. Tryptophan tryptophylquinone [TTQ]. C. Topaquinone.

TABLE 1. QUINOPROTEIN DEHYDROGENASES

Enzyme	Reference
Alcohol dehydrogenase	8
Aldehyde dehydrogenase	9
Aromatic amine dehydrogenase	10
Glucose dehydrogenase	11
Glycerol dehydrogenase	12
Lactate dehydrogenase	13
Methanol dehydrogenase	14
Methylamine dehydrogenase	15
Polyvinyl alcohol dehydrogenase	16
Polyethylene glycol dehydrogenase	17
Quinate dehydrogenase	18

review studies of the relevant properties and electron transfer reactions of this enzyme, and its relevance to biosensor technology.

Properties of Methylamine Dehydrogenase and Its Physiological Electron Acceptors

Methylamine dehydrogenase catalyzes the oxidation of methylamine to formaldehyde plus ammonia. The enzyme from *P. denitrificans* is very stable and exhibits no denaturation after incubation for 30 minutes at 70 degrees C or after incubation for 48 hours at pH values ranging from 3.0 to 10.0 (15). Similar stability has been reported for methylamine dehydrogenases from other sources as well.

The TTQ prosthetic group of methylamine dehydrogenase exhibits three redox states, oxidized, one electron reduced or semiquinone, and two electron fully reduced. The three redox forms can be distinguished spectroscopically, and each of these states is relatively stable under aerobic conditions (22). The oxidation- reduction midpoint potential $[E_m]$ value for the fully oxidized/fully reduced couple is +100 mV (22).

In vitro, methylamine dehydrogenase, and most quinoprotein dehydrogenases, are routinely assayed using artificial electron acceptors such as phenazine ethosulfate. The purification of this enzyme from *P. denitrificans* and the details of its assay with artificial electron acceptors have been described previously (23). During the oxidation of this enzyme in vivo, however, methylamine dehydrogenase donates electrons to a Type I "blue" copper protein, amicyanin (24), which further mediates electron transfer to a soluble cytochrome c-551i (25). As with methylamine dehydrogenase, the oxidized and reduced forms of amicyanin and cytochrome c-551i can be distinguished spectroscopically, which facilitates the study of their interactions. These three redox proteins, which form a soluble electron transfer chain that functions during methylamine oxidation by their host bacterium, have been purified and are well-characterized. Some relevant physical properties of these proteins are summarized in Table 2.

Kinetic Studies of Catalysis and Electron Transfer by Methylamine Dehydrogenase in Solution

To better understand the mechanisms of catalysis and electron transfer by methylamine dehydrogenase, an experimental system was devised by which these phenomena could be studied in solution under steady-state conditions. A kinetic approach was taken to describe this system in terms of apparent values of K_{cat}, and K_m for each of the components of this reaction. Methylamine, oxidized amicyanin, and oxidized cytochrome c-551i were treated as reactants for methylamine dehydrogenase. The initial rates of reaction were monitored by following the initial rate of reduction of cytochrome c-551i. The experimental details of this study have been described previously (3). From these data it was possible to obtain the relevant kinetic parameters for this

TABLE 2. PROPERTIES OF P. DENITIFICANS REDOX PROTEINS

Protein	Subunit Mol. Wt.	Subunit Structure	Redox Cofactor	E_m (mV)
Methylamine dehydrogenase	47,000			
	15,000	A_2B_2	quinone	+100
Amicyanin	15,000	monomer	copper	+296
Cytochrome c-551i	22,000	monomer	heme	+190

system. The K_{cat} for methylamine dependent cytochrome reduction was 1.1 x 103 min^{-1}. The apparent K_m values were 2.5 x 10^{-6}M for methylamine, 2.2 x 10^{-7}M for amicyanin, and 1.3 x 10^{-6}M for cytochrome c-551i. With this system it has been possible to assay for subtle perturbations in binding and activity due to environmental factors or specific modification of the individual proteins (3,27).

In general, electrostatic interactions are thought to play an important role in the proper orientation of proteins which allows intermolecular electron transfer to occur. It was of interest, therefore, to ascertain the effect of ionic strength on the methylamine-dependent reduction of cytochrome c-551i by these proteins. In the above mentioned assay, activity was optimal at low ionic strength and assays were routinely performed in 10 mM buffer. At saturating concentrations of methylamine and cytochrome c-551i, in the presence of added 0.2 M NaCl, the apparent K_m value for amicyanin increased from 2.2 x 10^{-7}M to 2.3 x 10^{-6}M. No change in the K_{cat} for this process was observed (3). These data are consistent with the suggestion that electrostatic interactions between amicyanin and methylamine dehydrogenase are involved in the proper orientation of the proteins which facilitates intermolecular electron transfer. This observation also supports the notion that this K_m value does in fact reflect a measure of the affinity of amicyanin for methylamine dehydrogenase.

It is of interest to compare the kinetic parameters of methylamine dehydrogenase in the presence of these physiological electron acceptors, with those parameters obtained using the artificial electron acceptors. Steady-state kinetic studies of this enzyme with artificial electron acceptors have been reported previously (28). Comparison of the observed values of $K_{cat}/K_{REOXIDANT}$ for the natural and artifical electron acceptors indicate that amicyanin is approximately 600-fold more efficient an electron acceptor for methylamine dehydrogenase than is PES.

Interactions between Methylamine Dehydrogenase and Amicyanin

Although methylamine dehydrogenase, amicyanin, and cytochrome c-551i from *P. denitrificans* are isolated as individual soluble proteins, they must form, at least transiently, a ternary complex in order to perform their physiologically relevant function (29). It was shown that direct electron transfer from reduced amicyanin to oxidized cytochrome c-551i does not occur in the absence of methylamine dehydrogenase. This has been explained in thermodynamic terms. The E_m values for cytochrome c-551i and amicyanin are +190 mV and +294 mV, respectively (26). Such a reaction is not, therefore, thermodynamically favorable. It was shown that amicyanin and methylamine dehydrogenase formed a weakly associated complex which caused a perturbation of the absorption spectrum of TTQ and a 73 mV shift in the redox potential of amicyanin to +221 mV (29). This raises the critical issue that the redox potential of a protein or enzyme may not be the same when it is in association with the surface of another macromolecule as it is when it is free in solution. This complex-dependent shift in potential which is described above is critical in that it facilitates the otherwise thermodynamically unfavorable electron transfer from amicyanin to the cytochrome. These complex-dependent phenomena were also only observed at low ionic strength.

The methylamine dehydrogenase-amicyanin complex and the interface between these redox proteins has been further characterized by chemical cross-linking (30), resonance Raman spectroscopy (31) and X-ray crystallographic studies (32). The data suggest that the stoichiometry of the complex is two amicyanin per methylamine dehydrogenase or one amicyanin per TTQ. The amicyanin is in contact with each of the dissimilar subunits of methylamine dehydrogenase. In contrast to the ionic strength dependence of the results described above, the chemical cross-linking and X-ray crystallographic studies indicate that the association between these proteins is stabilized primarily by hydrophobic interactions. A comparison of resonance Raman spectra of methylamine dehydrogenase and amicyanin free and in complex (31) revealed no significant changes, suggesting that neither the TTQ nor copper redox center is structurally altered during complex formation, and that the observed complex-dependent changes in their spectral and redox properties must be due to electrostatic interactions which accompany complex formation. Thus, efficient complex formation between these proteins must involve some combination of electrostatic and hydrophobic interactions.

The Mechanism of Intermolecular Electron Transfer from Proteins

Little information is available on weakly associated complexes of redox proteins. Most studies of multiprotein systems have involved species which are isolated as a tightly associated complex and which require harsh or denaturing conditions for resolution into their individual components. The study of multiprotein electron transport chains has been limited for the most part to membrane-bound systems. As it has been very difficult to crystallize these membrane proteins it has not been possible to obtain very much structural information to correlate with the available mechanistic data. The methylamine dehydrogenase-amicyanin complex is probably the best characterized complex of soluble weakly-associated redox proteins whose physiological function requires intermolecular electron transfer. It is the first such complex to be crystallized. The precise mechanism of the specific protein-protein association and the exact path of electron transfer between redox centers have not yet been completely determined. At this point in our analysis of the structure of the complex, however, certain details are known (33,34). The two indole rings which comprise the TTQ structure are not coplanar but at a dihedral angle of approximately 40-45 degrees. The o-quinone of TTQ is present in the active site, whereas the edge of the second indole ring, which does not contain the quinone is exposed on the surface of methylamine dehydrogenase (33). This portion of TTQ is within 10 angstrom of the copper atom of amicyanin, and one of the copper ligands of amicyanin, a histidine residue, appears to lie directly between the these redox centers (34). Thus, for the first time a picture is developing of a natural pathway through which long range electron transfer between proteins occurs.

Conclusions

The use of quinoproteins as biosensors is not purely theoretical. The construction of biosensors based upon the quinoproteins glucose dehydrogenase (36) and methanol dehydrogenase (37) have been reported. Direct electrochemistry of methylamine dehydrogenase has also been reported (38).

The field of the enzymology of quinoproteins is still quite young. As more progress is made towards the elucidation of quinoprotein structures and the understanding of the mechanisms of catalysis and electron transfer by protein-bound quinone cofactors, this knowledge may be useful in the logical design of specific quinoprotein-based electrodes. As additional quinoproteins are discovered and characterized, the potential range of applications for use of these enzymes in biosensors will increase as well.

Acknowledgments

Work performed in the author's laboratory has been supported by National Institutes of Health grant GM-41574.

Literature Cited

1. Duine, J. A.; Jongejan, J. A. Ann. Rev. Biochem. 1989, 58, 403- 426.
2. Davidson, V. L. Amer. Biotech. Lab. 1990, 8 (2), 32-34.
3. Davidson, V. L.; Jones, L. H. Anal. Chim. Acta 1991, 249, 235- 240.
4. Salisbury, S. A.; Forrest, H. A.; Cruse, W. B. T.; Kennard, O. Nature 1979, 280, 843-844.
5. McIntire, W. S.; Wemmer, D. E.; Christoserdov, A. Y.; Lidstrom, M. E. Science 1991, 252, 817-824.
6. Chen, L.; Mathews, F. S.; Davidson, V. L.; Huiginga, E.; Vellieux, F. M. D.; Duine, J. A.; Hol, W. G. J. FEBS Lett., 1991, 287, 163-166.
7. Janes, S. M.; Mu, D.; Wemmer, D.; Smith, A. J.; Kaur, S.; Mautry, D.; Burlingame, A. L.; Klinman, J. P. Science, 1990, 248, 981- 987.
8. Duine, J. A.; Frank, J.; Jongejan, J. A. Adv. Enzymol. 1987, 59, 169-212.
9. Ameyama, M.; Adachi, O. Methods Enzymol. 1982, 89, 491-497.
10. Nozaki, M. Methods Enzymol. 1987, 142, 650-655.
11. Duine, J. A.; Frank, J.; van Zeeland, J. K. FEBS. Lett. 1979, 108, 443-446.
12. Ameyama, M.; Shinagawa, E.; Matsushita, K.; Adachi, O. Agri. Biol. Chem. 1985, 49, 1001-1010.
13. Duine, J. A.; Frank, J. Trends Biochem. Sci. 1981, 6, 278-280.
14. Anthony, C. Adv. Microbial Physiol. 1986, 27, 113-210.
15. Husain. M.; Davidson, V. L. J. Bacteriol. 1987, 169, 1712-1717.

16. Shimoa, M.; Ninomiya, K.; Kuno, O.; Kato, N.; Sakazawa, C. Appl. Environ. Microbiol. 1986, 51, 268-275.

17. Kawai, F.; Yamanaka, H.; Ameyama, M.; Shinagawa, E.; Matsushita, K.; Adachi, O. Agri. Biol. Chem. 1985, 49, 1071-1076.

18. van Kleef, M. A. G.; Duine, J. A. Arch. Microbiol. 1988, 150, 132-136.

19. Anthony, C. In Bacterial Energy Transduction; Anthony, C., Ed.; Acedemic Press: San Diego, California, 1988; p. 293-316.

20. Husain, M.; Davidson, V. L.; Gray, K. A.; Knaff, D. B. Biochemistry 1987, 26, 4139-4143.

21. Davidson, V. L. Methods Enzymol. 1990, 188, 241-246.

22. Husain, M.; Davidson, V. L. J. Biol. Chem. 1985, 260, 14626-14629.

23. Husain, M.; Davidson, V. L. J. Biol. Chem. 1986, 261, 8577-8580.

24. Gray, K. A.; Knaff, D. B.; Husain, M.; Davidson, V. L. FEBS Lett. 1986, 207, 239-242.

25. Davidson, V. L.; Jones, L. H.; Kumar, M. A. Biochemistry 1990, 29, 10786-10791.

26. Davidson, V. L. Biochem. J. 1989, 261, 107-111.

27. Gray, K. A.; Davidson, V. L.; Knaff, D. B. J. Biol. Chem. 1988, 263, 13987-13990.

28. Kumar, M. A.; Davidson, V. L. Biochemistry 1990, 29, 5299-5304.

29. Backes, G.; Davidson, V. L.; Huitema, F.; Duine, J. A.; Sanders- Loehr, J. Biochemistry 1991, 30, 9201-9210.

30. Chen, L.; Lim, L. W.; Mathews, F. S.; Davidson, V. L.; Husain, M. J. Mol. Biol. 1988, 203, 1137-1138.

31. Chen, L.; Mathews, F. S.; Davidson, V. L.; Huizinga, E.; Vellieux, F. M. D.; and Hol, W. G. J. 1992, Proteins, in press.

32. Chen, L.; Durley, R.; Poloks, B. J.; Hamada, K.; Chen, Z.; Mathews, F. S.; Davidson, V. L.; Satow, Y.; Huizinga, E.; Vellieux, F. M. D.; and Hol, W. G. J. 1992, Biochemistry, in press.

33. D'Costa, E. J.; Higgins, I. J.; Turner, A. P. F. Biosensors 1986, 2, 71-87.

34. Zhao, S.; Lennox, R. B. Anal. Chem. 1991, 63, 1174-1178.

35. Burrows, A. L.; Hill, H. A. O.; Leese. T. A.; McIntire, W. S.; Nakayama, H.; Sanghera, G. S. Eur. J. Biochem. 1991, 199, 73-78.

RECEIVED May 27, 1992

Chapter 2

Microbiosensors for Food Analysis

Isao Karube and Masayasu Suzuki

Research Center for Advanced Science and Technology, University
of Tokyo, Komaba, Meguro-ku, Tokyo 153, Japan

Several examples of microbiosensors for food analysis,
which were developed by our group, are described.
Microbiosensors for saccharides, alcohols, amino acids
and nucleotides were constructed with immobilized
biocatalysts and micro transducers, such as ion sensitive
field effect transistors, amorphous silicon ion sensitive
field effect transistors, micro oxygen electrodes, micro
hydrogen peroxide electrodes and platinized carbon fiber
electrodes. By the use of microbiosensor technology,
several biosensors can be easily integrated in one chip.
Such a sensor is very useful for the simultaneous
determination of multiple components of food.
Furthermore, disposable type microbiosensors are very
useful for on-site monitoring of food processes.

The determination of organic compounds is very important in food industries.
However, conventional methods often require a long reaction time and
complicated procedures. Biosensors employing immobilized biocatalysts have
definite advantages, since biosensors show excellent selectivity for biological
substances and can directly determine a single compound in a complicated
mixture without need for a prior separation step (1-3). Enzyme reactions
consume the chemical substances and result in the formation of products
which can be measured by electrodes and other devices. The enzymes and
these devices may be combined to produce a biosensor with extremely good
selectivity. In actuality, several biosensors such as a glucose sensor, lactate
sensor, alcohol sensor and acetic acid sensor are used in food industry
processes.

Recently, since progress has been made in semiconductor fabrication
technology, various kinds of small-size sensing devices have been employed for
the construction of biosensors. These biosensors are referred to as "
microbiosensors" (4-6). Microbiosensors have many advantages over the
conventional biosensors as mentioned in the next section, which may be
applied to food industries.

Determination of several compounds at the same time is quite

0097–6156/92/0511–0010$06.00/0

important in the food analysis. For example, inosine, inosine monophosphate and hypoxanthine are used for estimating the freshness of fish. Also for the estimation of taste, composition rate of amino acids and nucleotides is very important.

By the use of microbiosensor technolology, several biosensors can be easily integrated in one chip. Such an integrated biosensor is very useful for the determination of multiple components of food.

In case of microbiosensors, mass production is also possible and disposable-type biosensors can be fabricated, such that quantitative biosensors can be used as "test paper strip". This is extremely useful for on-site monitoring of food processes.

In this paper, several microbiosensors for food industries which were developed by our group are described.

Microbiosensors

Conventional biosensors are composed of a detector, and immobilized biocatalyst. Oxygen, hydrogen peroxide and pH electrodes have been utilized as electrochemical biosensors. However, with the increase in demand for biosensor miniaturization, ion sensitive field effect transistors (ISFETs) and other micro electrodes which are produced by semiconductor fabrication technology, are being utilized instead of common electrodes. The miniaturization of biosensors has definite advantages compared with conventional biosensors. First, only a tiny amount of sample is required for the measurement. Second, implantation of the biosensor is possible, which enables *in vivo* measurement. Third, mass production of such microbiosensors is possible, thus realizing disposable type biosensors. Fourth, various microbiosensors can be integrated, in order to construct multi-functional integrated microbiosensors.

The ion sensitive field effect transistor (7) is a micro pH sensing device made by using silicon fabrication technology, and is attractive because of its small size and mass producibility. It has found use as a micro-ion-sensitive device in proton, sodium and potassium ion and surface charge measurements. The first micro biosensor was constructed using ISFETs (8).

The conventional ISFET device can only be manufactured by using a silicon wafer as the substrate. Recently, devices made with amorphous silicon have received much attention. Various substrates such as glass, plastics, etc., can be used for devices made with amorphous silicon and transistors can be manufactured with a number of different structures. The authors have developed an amorphous silicon ISFET (a-ISFET)(9). Fig.1 shows the structure of the a-ISFET. Using the a-ISFET as a transducer, sensors were developed for glucose (10), hypoxanthine (11) and inosine (12).

Many enzyme and microbial sensors have been developed utilizing an oxygen electrode as a transducer. Therefore the development of micro oxygen electrodes will greatly contribute to the miniaturization of biosensors. The authors have developed an improved micro oxygen electrode based on conventional semiconductor fabrication technology (13). The structure of the micro oxygen electrode is shown in Fig.2. The structure has a V-formed groove, 300 μm deep, and two gold electrodes over a SiO_2 layer which electrically insulates them. Each gold electrode covers about half of the oxygen electrode.

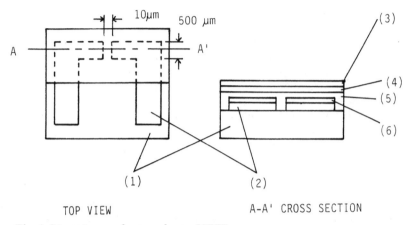

Fig.1 Structure of amorphous ISFET
(1)Glass plate; (2)Aluminium electrode; (3)SiO layer;
(4)a-SiNx layer; (5)a-Si:H layer; (6)n+ layer
(Reproduced with permission from ref.9. Copyright 1989 Elsevier Sequoia S.A.)

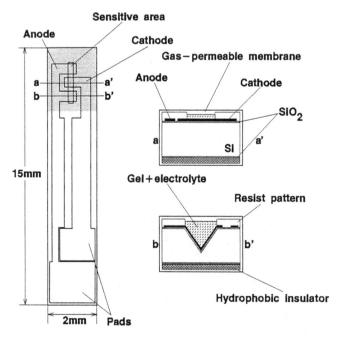

Fig.2 Structure of micro oxygen electrode

Calcium alginate gels containing a 0.1M potassium chloride aqueous solution as an electrolyte were poured into the groove and then covered by the gas permeable membrane. The calcium alginate layer was approximately 300 μm thick, and the gas permeable membrane was 2 μm thick. Using the micro oxygen electrode, a glucose sensor (13), hypoxanthine sensor (14), microbial CO_2 sensor (15) and hybrid L-lysine sensor (16) were constructed.

An enzyme sensor based on oxidase reactions can be constructed using an oxygen electrode or an H_2O_2 electrode. The authors have developed a micro H_2O_2 electrode (17). The structure of the micro H_2O_2 electrode is shown in Fig.3. A micro gold electrode was prepared on silicon nitride, using the vapor deposition method. Part of the gold electrode was coated with Ta_2O_5 for electrical insulation. This electrode works as an H_2O_2 sensor when the potential between both gold electrodes is set at 1.1 V. A glucose sensor, monoamine sensor and putrescine sensor were developed using this electrode.

Recently the authors have also focused on the development of "ultra"-micro biosensors using carbon fiber electrodes (18,19). Carbon fibers with 7 μm diameter were used to construct ultra-micro biosensor. Microcomputer-controlled potentiometric polarization techniques have been applied to analyze hydrogen peroxide using carbon fiber electrodes. Ultra micro biosensors for acetylcholine, glucose and glutamic acid were constructed by immobilizing acetylcholine esterase/choline oxidase, glucose oxidase and glutamate oxidase (19) on the carbon fiber electrode.

Microbiosensors for Saccharides

In the field of food science, analysis of sugars is very important. For example, glucose, lactose and sucrose are contained in various kinds of foods, and the determination of those sugars may be important in the control of processing and quality. In practice, enzyme sensors are used for the determination of glucose in fruit juice, wine, sake and soy sause.

Glucose oxidase (GOD) is used for glucose sensor using the following reaction:

$$GOD$$
$$Glucose + H_2O + O_2 \text{ ------> } Gluconolactone + H_2O_2$$

By using an oxygen or hydrogen peroxide sensor, a glucose sensor can be constructed. A micro glucose-sensor was constructed by the use of a micro-H_2O_2 electrode developed utilizing integrated circuit technology (17). The procedure for glucose oxidase immobilization onto the micro electrode is performed   as follows. Approximately 100 μl of gamma-aminopropyl-triethoxysilane was vapourized at 80 C, 0.5 Torr, for 30 min onto the electrode surface, followed by 100 μl of 50 % glutaraldehyde vapourized under the same conditions. The modified micro electrodes were then immersed in enzyme solution containing bovine serum albumin and glutaraldehyde, the enzyme becoming chemically bound to the surface of the micro electrode by schiff-

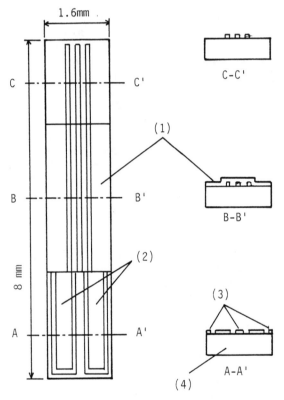

Fig.3 Structure of micro hydrogen peroxide electrode
(1)Photoresist; (2)Working electrode; (3)Counter electrode;
(4)Sapphire
(Reproduced with permission from ref.17. Copyright 1986 Marcel
Dekker, Inc.)

linkage. Fig.4 shows the typical response curve of the micro glucose sensor. The output current increased after injection of a sample solution, steady state being reached within 5 min.

A linear relationship was observed between the current increase and glucose concentration in the range 0.1 to 10 mg/dl glucose. Examination of the selectivity of the micro-glucose sensor indicated no response to other compounds such as galactose, mannose, fructose and maltose. Therefore, the selectivity of this sensor for glucose is highly satisfactory. Continuous operation of the sensor in 10 mg/dl glucose produced a constant current output for more than 15 days and 150 assays. Thus, this micro-glucose sensor possesses both selectivity and good stability, its potential use as a micro glucose sensor being very good.

By the use of micro electrodes, several kinds of microbiosensors can be integrated onto a chip. The authors developed an integrated glucose and galactose sensor (20). Glucose oxidation via quinoprotein glucose dehydrogenase (GDH) occurs by the electron transfer between glucose and pyrrolequinoline quinone (PQQ) in the GDH. This electron tansfer is not affected by dissolved oxygen. A glucose sensor was constructed by utilizing this GDH reaction. Quinoprotein GDH and galactose oxidase (GAOD) was immobilized onto a microelectrode. The concentration of glucose was determined by using an electron mediator, and that of galactose by measuring dissolved oxygen. The utilized quinoprotein GDH was purified from *Pseudomonas fluorescens*, which was kindly provided by Prof. Ameyama (Yamaguchi University, Japan). The integrated microbiosensor was constructed by casting a GDH, GAOD and BSA solution onto a microelectrode, and crosslinking with glutaraldehyde. In this experiment, ferrocene monocarbonate (FCA) was dissolved in the reaction buffer, and utilized as an artificial electron mediator. The terminal voltage of the sensor measuring FCA was set at +0.35 mV and that of the sensor measuring dissolved oxygen was set at -0.3 mV versus Ag/AgCl. The determination of galactose concentration by the sensor was first attempted. Since the GAOD reaction is also catalyzed by electron transfer to FCA, both electrodes responded to galactose. Fig.5 shows the calibration curves for glucose in the presence of several concentrations of galactose. The indicated plots were obtained by the injection of glucose after the steady current was observed in the presence of galactose. As can be seen from Fig.5, with the increase of galactose concentration, the response to glucose decreased. However, by comparing the glucose response to galactose response, as monitored by the dissolved oxygen content, an appropriate calibration curve for glucose can be achieved. Therefore, the simultaneous measurement of glucose and galactose can be performed by utilizing this integrated microbiosensor.

Microbiosensors for Alcohols

For on-line monitoring of fermentation processes, the small and selective ethanol sensor is applicable. Kitagawa *et al.* developed a micro alcohol sensor

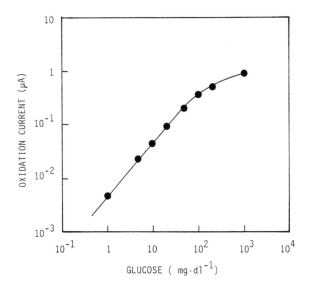

Fig.4 Relationship between glucose concentration and
oxidation current
(Reproduced with permission from ref.17. Copyright 1986 Marcel
Dekker, Inc.)

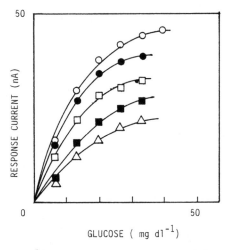

Fig.5 Dependence of the response of the oxidizing mediators
electrode to glucose on galactose concentration.
    The graphs were obtained on injection of glucose after
a steady-state current was observed in the presence of galactose.
Galactose concentration:($\bigcirc$)0; ($\bullet$)16.6; ($\square$)33.1; ($\blacksquare$)49.5;
($\triangle$)65.8 mg·dl^{-1}.
(Reproduced with permission from ref.20. Copyright 1989 Elsevier
Sci. Pub.)

consisting of an immobilized acetic bacteria, a gas permeable membrane and an ISFET (21). This sensor was utilized for the determination of ethanol.

Acetic acid bacteria have been widely used in industry to make vinegar and ascorbic acid by oxidative fermentation. In acetic acid fermentation, both alcohol dehydrogenase (ADH) and aldehyde dehydrogenase (ALDH) are involved in oxidization of ethanol to acetic acid via acetaldehyde (Fig.6). When acetic acid bacteria are immobilized on an ISFET, ethanol is oxidized to acetic acid by those bacteria, causing a pH change to occur at the surface of ISFET. Ethanol concentration can thus be measured by combining an ISFET and acetic acid bacteria.

Acetobacter aceti IAM 1802 was used for the micro alcohol sensor. The microorganisms were immobilized on the gate surface of ISFET, in calcium alginate gel. The microorganisms-immobilized ISFET and wire-like Ag-AgCl electrode were placed in a small shell (6x3x22mm) that had a gas permeable membrane (a porous Teflon membrane, 0.5 μm pore size) fitted to the side. The inside of the shell was filled with inner buffer solution (5mM Tris-HCl buffer containing $0.1M$ CaC^{l2}, pH 7.0).

The system consisted of a thermostated circulating jacketed vessel (3ml volume), the microbial-FET alcohol sensor, the circuit for measurement of the gate output voltage (Vg), an electrometer and a recorder. The initial rate change of gate voltage with time, dV/dt , was plotted against the logarithmic value of the ethanol concentration. The minimum detectable response to ethanol was obtained at approximately 0.1 mM. A linear relationship was observed over the range between 3 - 70 mM (Fig.7).

The response of the micro alcohol sensor was stable over a wide pH range (pH 2 - 12), however, at pH's lower than 6, the sensor responded to acetic acid. Volatile organic acids such as acetic acid do not dissociate in lower pH solution, and so, penetrate through the gas permeable membrane. At pH's higher than 6, the volatile organic acids dissociate to the individual ions, and do not interfere with the measurement of ethanol. The sensor was stable for 15 h, at 15°C, and demonstrated its usefulness for the determination of ethanol, especially in terms of its selectivity and potential for miniaturization. It is expected that this sensor can be integrated into a multi-functional sensor capable of simultaneously determining multiple substrates in a solution of complex composition.

Microbiosensors for Amino Acids

The detection of amino acids is important in the quality control of several types of food. Glutamic acid is often used in Japanese or Chinese food. Especially in the soup producing processes, monitoring of glutamate concentration is quite important.

A micro glutamate sensor was developed using glutamate oxidase and a platinized carbon fiber disk (PCD) electrode (19).

Glutamate oxidase (Yamasa Shoyu Co. Ltd) and bovine serum albumin were mixed in the ratio of 1:3, and mixed with distilled water and

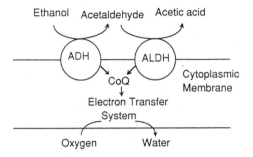

Fig.6 Ethanol oxidation system of Gluconobactor.
ADH:alcohol dehydrogenase; ALDH:aldehyde dehydrogenase;
CoQ:coenzyme Q.
(Reproduced with permission from ref.21. Copyright 1988 Elsevier
Sci. Pub.)

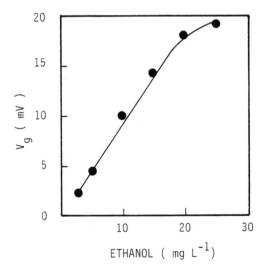

Fig.7 Calibration graph for ethanol
(Reproduced with permission from ref.21. Copyright 1988 Elsevier
Sci. Pub.)

photocrosslinking polyvinyl alcohol (PVA-SbQ, Toyo Gosei Co. Ltd) in the ratio of 1:20:20. The mixture was attached to the tip of the electrode, and was left for 10 min under a fluorescence lamp.

The electrode was dipped into phosphate buffer and left for 3 min in a glutaraldehyde atmosphere. Current response of this ultra micro glutamate sensor, using a PCD electrode, to stepwise addition of glutamate was shown in Fig.8. Silver-silver chloride was used as reference electrode. Responses were very stable and response time was within 12 sec. Fig.9 shows the calibration curve for glutamate. Calibration range of the ultra micro glutamate sensor was from 2 μM to 12 mM. Standard deviation at 2 μM, obtained from 60 repeat measurements was 0.498 pA. Since the actual response current was 8.498 pA, error for 2 μM glutamate measurement was 5.9%.

In making L-amino acid sensors, carboxylases are often used in the receptor area because these enzymes are readily available. To develop amperometric L-amino acid sensors, amperometric bactrial CO_2 sensors were developed with a conventional galvanic oxygen electrode and autotrophic bacteria and used in an amperometric L-tyrosine sensor (22). In addition, a disposable, miniature Clark-type oxygen electrode was developed utilizing microfabrication techniques (13). Furthermore, a bacterial CO_2 sensor was made using a miniature oxygen electrode (15). The CO_2 sensor can be used in the miniature L-amino acid sensors. The authors developed a hybrid L-lysine sensor consisting of an immobilized L-lysine decarboxylase and a miniature bacterial CO_2 sensor (16). The bacteria were immobilized in a calcium alginate gel in a miniature oxygen electrode cell together with the electrolyte. The enzyme was immobilized in a bovine serum albumin matrix on a gas-permeable membrane (Fig.10). The cell was formed on a silicon substrate by anisotropic etching and had a two-gold electrode configuration. The response time of the L-lysine sensor was 1-3 min. The optimum pH was 6.0 and the optimum temperature was 33°C. The response to L-lysine concentration was linear from 25 to 400 μM. Reproducible responses were obtained by adding more than 1 uM pyridoxal-5'-phosphate. The sensor had excellent selectivity for L-lysine and a stable response for more than 25 repetitive operations.

Microbiosensors for Nucleotides and Application for Freshness Estimation

Determination of nucleotide-related compounds is very important in food technology, especially for freshness estimation of fish or meat, and taste sensing. The determination of ATP (adenosine triphosphate) is important in fermentation processes. A micro ATP sensor was developed using H+-ATPase and an ISFET (23). H^+-ATPase (EC 3.6.1.3) in biological membranes catalyses the production or hydrolysis of ATP. Furthermore, the enzyme has many functions, such as proton transport, which could be utilized by a bio-molecular device. H^+-ATPase was prepared from a thermophilic bacterium PS3 and is classified as thermophilic F1 (TF1) ATPase. The procedures employed in constructing the ATP sensor and measurements of gate voltage were described elsewhere (23). The differential gate output voltage reached steady state

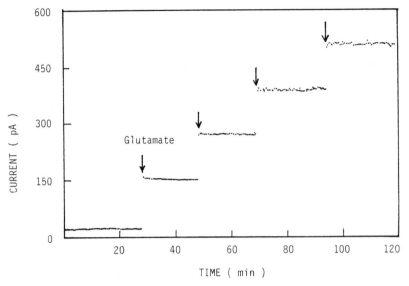

Fig.8 Typical response of glutamate sensor to a stepwise addition
of glutamate. An arrow shows injection of 25µl of 100mM glutamate
solution into 50 ml phosphate buffer(pH 7.0, 0.1M).
(Reproduced with permission from ref.19. Copyright 1991 IEEE)

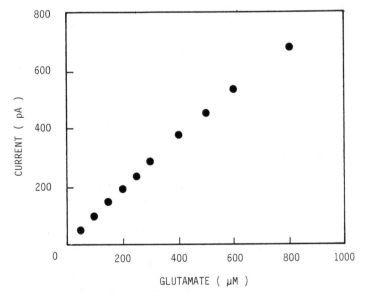

Fig.9 Calibration curve of the glutamate sensor
Glutamate concentration range: 50 to 800 µM.
(Reproduced with permission from ref.19. Copyright 1991 IEEE)

approximately 4-5 minutes after injection of ATP. The initial rate of the differential gate output voltage change after injection of ATP was plotted against the logarithm of ATP concentration. Fig. 11 shows a calibration curve of the ATP sensor system.

A linear relationship was obtained between the initial rate of the voltage change and the logarithm of ATP concentration over the range 0.2 to 1.0 mM ATP. Slight responses were obtained when glucose, urea, and creatinine were applied to the system. The response of the system to ATP was retained for 18 days.

A sensor for the determination of inosine was also prepared by a combination of the enzyme system (shown below) and an amorphous silicon ISFET (12).

nucleoside phosphorylase
Inosine + Pi ------------------------> Hypoxanthine + Ribose-P
xanthine oxidase
Hypoxanthine + 2O$_2$ ----------------> Uric acid + 2H$_2$O$_2$

Nucleoside phosphorylase and xanthine oxidase were covalently immobilized on polyvinylbutyral membrane containing 1,8-diamino-4-aminomethyloctane. The optimum conditions for determination of inosine were pH 7.5, 32 C. The sensor gave a linear relationship between the initial rate of the output gate voltage change and the logarithm of inosine concentration between 0.02 and 0.1 mM. Determination of inosine was possible within 7 min. The system could be used for two weeks with about 35% loss of enzymatic activity.

Hypoxanthine, one of the intermediates of autolysis process in fish or meat, accumulates; its concentration increases with prolonged storage, and can be used as an indicator of fish freshness. The authors developed a disposable type micro enzyme sensor using a Clark-type micro oxygen electrode and immobilized xanthine oxidase (EC 1.1.3.22) for the simple assay of hypoxanthine (14). The sensor showed a good response to hypoxanthine and allowed the determination of hypoxanthine in the concentration range between 6.7 - 180 μM (Fig.12).

Similarly, a micro hypoxanthine sensor could be constructed with immobilized xanthine oxidase and a micro electrode, such as an ISFET, amorphous silicon ISFET (11) and micro hydrogen oxide electrode.

The determination of fish freshness is very important in the food industry for the manufacture of high quality products. Indication of fish freshness, such as nucleotides, ammonia, amines, volatile acids, catalase activity and pH value have been proposed. Among those indicators, nucleotides produced by ATP decomposition are considered the most reliable and useful indicator. The degradation pathway of ATP in fish muscle is summarized in Fig.13. After the death of the fish, the ATP and ADP decompose rapidly, and inosine-5'-mono-phosphate (IMP) is formed. Changes in the AMP (Adenosine mono- phosphate) level are negligible, and the amount of AMP is also small. Inosine (HxR) and hypoxanthine (Hx) gradually accumulate with decomposing

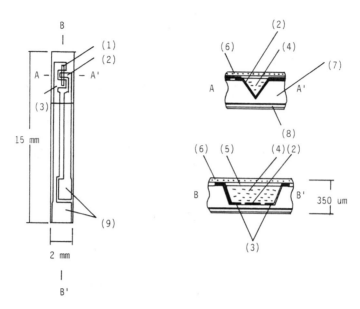

Fig.10 Structure of L-lysine sensor
(1)Sensitized area; (2)cathode; (3)anode; (4)calcium alginate gel
containing bacteria and electrolyte; (5)gas-permeable membrane;
(6)immobilized enzyme;(7)silicon substrate; (8)hydrophobic insulator;
(9)pad.(Reproduced with permission from ref.16. Copyright 1990
Elsevier Sci.Pub.)

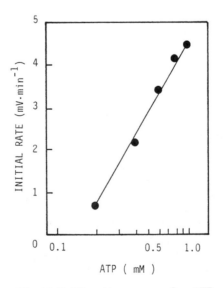

Fig.11 Calibration curve for ATP

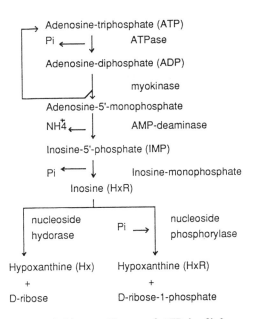

Fig.13 Degradation pathway of ATP in fish muscle

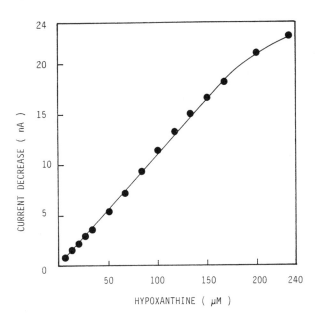

Fig.12 Calibration curve for hypoxanthine
Temperature:37°C; pH 7.3, 0.1M potassium phosphate buffer.
(Reproduced with permission from ref.14. Copyright 1989 Marcel
Dekker, Inc.)

IMP, because the degradations of HxR and Hx are rate-determining steps in this pathway. Therefore, KI value presented by the following equation, was used as an indicator of fish freshness (24).

$$KI = 100x([HxR] + [Hx])/([IMP] + [HxR] + [Hx])$$

By the integration of three microbiosensors for IMP, inosine and hypoxanthine in one chip, fish freshness might be able to be determined with the integrated micro enzyme sensor.

Conclusion

Several examples of microbiosensors for food industries, which were developed by our group, were described. Various kinds of biosensors, especially enzyme sensors, have already been used for food analysis. Microbiosensors described in this article might be practically used in food industry within a short time. In the future, food analysis may focus on taste, smell or freshness sensing, rather than on only a single component determination. The integrated microbiosensor could become a key technology for the realization of those sensing requirements.

Literature Cited

1. Turner,A.P.F.;Karube,I.;Wilson,G.*Biosensors-Fundamental and Applications*;Oxford University Press:Oxford,1987.
2. Mosbach,K.*Method in Enzymology*;Academic Press:London and New York,1988,Vol.137.
3. Cass,A.E.G.*Biosensors-A Practical Approach*;IRL Press: Oxford,1990.
4. Karube,I.;Sode,K.*Swiss Biotech.* **1989**,7,25-32.
5. Karube,I.;Sode,K.;Tamiya,E.*J.Biotechnol.* **1990**,15,267-282.
6. Karube,I.*Makromol. Chem.,Makromol. Symp.* **1988**,17,419-427.
7. Matsuo,T.;Wise,K.D.*IEEE Trans. on BME.* **1974**,BME-21,485.
8. Caras,S.;Janata,J.*Anal. Chem.* 1980,52,1935.
9. Gotoh,M.;Oda,S.;Shimizu,I.;Seki,A.;Tamiya,E.;Karube,I. *Sensors & Actuators.***1989**,16,55-65.
10.Gotoh,M.;Tamiya,E.;Seki,A.;Shimizu,I.;Karube,I.*Anal.Lett.***1989**, 22, 309-322.
11.Tamiya,E.;Seki,A.;Karube,I.;Gotoh,M.;Shimizu,I.*Anal. Chim.Acta.* **1988**, 215,301-305.
12.Gotoh,M.;Tamiya,E.;Seki,A.;Shimizu,I.;Karube,I.*Anal.Lett.***1988**, 21,1785-1800.
13.Suzuki,H.;Tamiya,E.;Karube,I.*Anal.Chem.* **1988**,60,1078-1080.
14.Suzuki,M.;Suzuki,H.;Karube,I;Schmid,R.D.*Anal.Lett.* **1989**,22, 2915-2927.

15. Suzuki,H.;Kojima,N.;Sugama,A.;Takei,F.;Ikegami,K.;Tamiya,E.; Karube,I. *Electroanalysis*. **1989**,1,305-309.

16. Suzuki,H.;Tamiya,E.;Karube,I.*Anal.Chim.Acta*. **1990**,229,197-203.

17. Murakami,T.;Nakamoto,S.;Kimura,J.;Kuriyama,T.;Karube,I.*Anal. Lett*.**1986**, 19,1973-1986.

18. Tamiya,E.;Sugiura,Y.;Akiyama,A.;Karube,I.*Ann.N.Y.Acad.Sci.* **1990**, 613, 396-400.

19. Tamiya,E.;Sugiura,Y.;Akiyama,A.;Suzuki,M.;Karube,I.*Tech.Digest of Transducers'91*;IEEE:New York, **1991**; pp 381-384.

20. Yokoyama,K.;Sode,K.;Tamiya,E.;Karube,I.*Anal.Chim.Acta*.**1989**, 218,137- 142.

21. Tamiya,E.;Karube,I.;Kitagawa,Y.;Ameyama,M.;Nakashima,K.; *Anal.Chim.Acta.* **1988**,207,77-84.

22. Suzuki,H.;Tamiya,E.;Karube,I.*Anal.Lett.***1989**, 22,15-24.

23. Gotoh,M.;Tamiya,E.;Karube,I.;Kagawa,Y.*Anal.Chim.Acta*.**1986**, 187,287-291.

24. Karube,I.;Matsuoka,H.;Suzuki,S.;Watanabe,E.;Toyama,K.J.*Agric. Food Chem*.**1984**,32,314-319.

RECEIVED May 27, 1992

Chapter 3

Biosensors for Food Analysis

A. A. Suleiman[1,3] and G. G. Guilbault[2,3]

[1]Department of Chemistry, Southern University at Baton Rouge,
Baton Rouge, LA 70813
[2]Department of Chemistry, University of New Orleans,
New Orleans, LA 70138

There has been a considerable interest in developing new assays for food analysis to meet the demands of the different regulatory health agencies. Biosensors offer a new alternative due to their inherent specificity, simplicity and quick response. Selected biosensors which have been developed recently in our laboratory featuring applications in food analysis are discussed. These include enzyme electrodes and fiber optic probes for the detection and determination of various substances (fructose, glutamate, aspartame, hydrogen peroxide, glucose and sulfite).

Enzyme Electrodes Based Biosensors

A Simplified Fructose Biosensor. Several procedures have been described for the selective determination of D-fructose where most of the enzymatic assays are based on the following sequential enzymatic reactions:

$$\text{Fructose} + \text{ATP} \xrightarrow{\text{Hexokinase}} \text{fructose 6-phosphate} + \text{ADP}$$

$$\text{Fructose 6-phosphate} \xrightarrow{\text{Phosphoglucose}} $$

$$\text{glucose 6-phosphate (G-6-P)}$$

$$\text{Glucose 6-phosphate} + \text{NADP}^+ \xrightarrow{\text{G-6-DPH}} $$

$$\text{NADPH} + \text{H}^+\text{-6-phospho-}\delta\text{-lactone}$$

[3]Current address: Universal Sensors, Inc., 5258 Veterans Boulevard, Suite D, Metairie, LA 70006

0097–6156/92/0511–0026$06.00/0

The determination of fructose is usually accomplished by spectrophotometric (1), fluorometric (2) and electrochemical (3) measurements of the produced NADPH. The major disadvantages of the above mentioned methods are the high cost due to the expensive enzymes and co-enzymes consumed in the assay and the potential interference of glucose.

Fructose was also determined colorimetrically using soluble D-fructose 5-dehydrogenase (FDH) in the presence of a mediator (4,5). Additionally, an amperometric flow injection technique was reported based on the following enzymatic reaction (6,7):

```
                              FDH
D-fructose + 2 ferricyanide ---->

                    5-keto-D-fructose + 2 ferrocyanide
```

Although, a low detection limit and a wide range were both achieved, the reactor lost about 50% of its initial activity in seven days.

The proposed enzyme electrode is based on the same principle and was described earlier (8). It was constructed by immobilizing FDH on selected permeable membranes using the cross-linking method via glutaraldehyde and bovine serum albumin (BSA) matrix. The immobilized enzyme membrane was mounted on a combined electrode composed of a Pt working electrode and an Ag/AgCl reference electrode then secured with a dialysis membrane. The change in current due to the oxidation of $Fe(CN)_6^{-4}$ produced during the enzymatic reaction was measured at an applied potential of 0.385 and correlated to the concentration of fructose.

Membrane Performance. The permeability of $Fe(CN)_6^{-4}$ through different membranes was evaluated to select the suitable membrane in order to obtain the best sensitivity and fastest response. The change of current due solely to the amount of $Fe(CN)_6^{-4}$ which diffused through the membrane and oxidized to $Fe(CN)_6^{-3}$ was measured under identical conditions of stirring rate, temperature, pH, etc., which affect the diffusion rate. Among the membranes tested, nuclepore (0.1 μm and 1 μm pore size) exhibited the best permeability (Table I).

**Table I.   Permeability of $Fe(CN)_6^{-4}$
Through Different Membranes**

Membranes		Time to Reach Steady State	Steady State Current (nA)
1.	Immobilon	2 min	57.8
2.	Nuclepore (1μm)[a]	15 s	67.8
3.	Nuclepore (0.03 μm)	30 s	40.4
4.	Nuclepore (0.1 μm)	15 s	69.3

[a]Pore size

Considering the stability of the immobilized enzyme electrode, the small pore-size (0.1 μm) nuclepore membrane was eventually used to prevent the penetration of leached immobilized enzyme molecules, thereby minimizing their deposition onto the electrode surface.

Analytical Characteristics and Applications. A typical calibration curve constructed using optimum conditions (pH = 5, $[Fe(CN)_6^{-3}]$ = 2 mM) was linear in the concentration range 1.00×10^{-5} - 1.00×10^{-3}M fructose. The electrode retained 90% of its initial activity after 20 days. However, the activity declined to about 75% after one month, and leveled to 60% after two months. During a two month period, more than 300 assays were performed using one electrode.

The immobilzed FDH enzyme electrode was used successfully for the determination of fructose in several commercial products (Table II).

Table II. Comparison Study of the Present Method with the AOAC Standard Method

Sample	Fructose (%) Proposed Method	AOAC Method
1. Apple juice[a]	20.5	20.2
2. Orange juice[b]	5.80	6.40
3. Orange juice[c]	2.92	2.69
4. Apple-Cherry juice[c]	5.4	5.24
5. Apple-Grape juice[c]	5.86	5.94
6. Apple juice[c]	6.34	6.87
7. Pear juice[c]	5.24	5.19

[a]Seneca concentrated
[b]Minute Maid concentrated
[c]Gerber baby food

Glutamate Enzyme Electrode. There is a considerable interest in the rapid determination of glutamate, which is found in a variety of foods and biological materials. Glutamate is a potent neuroexcitatory amino acid associated with certain behavior patterns such as aggressive behavior, retarded learning, morphine-induced muscular rigidity and retrograde amnesia. In addition, the popular flavor enhansor Monosodium glutamate (MSG) has been linked to the Chinese Restaurant Syndrome (9). Potentiometric enzyme electrodes for glutamate assay using glutamate decarboxylase and glutamate dehydrogenase have been reported, however, these proved to be unstable. An amperometric electrode based on immobilized L-glutamate oxidase was also reported (10), but this electrode suffered from interferences by L-glutamine, L-aspartate and L-asparagine. Our enzyme electrode is based on the immobilization of L-glutamate oxidase on pre-activated Immobilon-AV Affinity membranes. The reaction catalyzed by glutamate oxidase is given by

$$\text{L-glutamate} + O_2 + H_2O \xrightarrow{\text{L-glutamate oxidase}} \alpha\text{-ketoglutarate} + NH_4^+ + H_2O_2$$

Oxygen or hydrogen peroxide may be monitored amperometrically, and the current change is directly proportional to glutamate concentration. To eliminate interferences from electro-oxidizable species, a hydrophobic oxygen membrane or a size-exclusion membrane (100 M.W.-cut-off membrane) was inserted between the electrode and the enzyme membrane. Chemical amplification of the response was achieved by co-immobilizing L-glutamate dehydrogenase which catalyzes the following reaction:

$$\text{L-glutamate} + NAD(P)^+ + H_2O \overset{\text{glutamate}}{\underset{}{\xrightarrow{\text{dehydrogenase}}}} \alpha\text{-ketoglutarate} + NH_4^+ + NAD(P)H$$

Ammonium chloride and NAD(P)H were added to the buffer solution to drive the reaction towards production of more L-glutamate. The two-enzyme system causes cycling of L-glutamate, α-ketoglutarate, and NH_4^+ between the two enzymatic reactions.

Electrode Preparation. The Immobilon-AV Affinity Membrane, 0.65 μm pore size (Millipore Corporation, Bedford, MA) was attached to an inverted electrode jacket with a rubber ring. 5 μl of 72 units/ml of L-glutamate oxidase solution and 5 μl of 400 units/ml glutamate dehydrogenase solution were pipetted onto the membrane and allowed to dry at room temperature for 1-2 hr and were kept immersed in Dulbecco's buffer at 4° when not in use. The buffer and filling solution was 0.1 M KH_2PO_4, 0.1 M NaCl, pH 7.0. A constant potential of +650 or -650 mV vs Ag/AgCl was applied to the Pt working electrode for the peroxide and oxygen sensor, respectively. An O_2 hydrophobic membrane was inserted between the enzyme membrane and the electrode jacket if the oxygen consumption was measured. Steady state currents were measured with a BAS LC-4B Amperometric Detector (BioAnalytical Systems, Inc., West Lafayette, IN). The electrode was also included in a flow injection system for continuous assay. More details about the experimental set-up and procedure can be found elsewhere (11,12).

Optimization of Parameters. The performance of the electrode was evaluated to determine the optimum conditions such as: enzyme loading, membrane type, drying time and pH. The immobilization on Immobilon produced the best electrode compared with other membranes (Pall Immunodyne Immuno-Affinity membrane, UltraBind Affinity membranes and Nuclepore membranes). The effects of enzyme loadings and immobilization time were eventually studied for an electrode constructed using Immobilon. The response increased with enzyme load up to 0.30 and 2.0 units of glutamate oxidase and glutamate dehydrogenase, respectively, per 0.78 cm^2 (jacket tip area) of membrane and up to 60 min. immobilization time. The effect of pH on the response for four buffer systems (acetate, phosphate, tris, carbonate) with pH ranging from 3.70 to 11.0 was studied. It was concluded that the response is optimum at pH 6.8 and does not vary much with the type of buffer used. The stability study of the enzyme membrane showed that the best storage condition is in Dulbecco's buffer at 4°, where the membrane lost 25% of its initial activity after 7 days, but was then stable over a period of 320 days and at least for 200 assays. It was also evident that the stability under operating conditions depends on the sample matrix. In buffered solutions of glutamate and other amino acids, the enzyme membrane remains active for a long time, however, matrices containing reducing agents or heavy metals deactivate the enzyme.

Analytical Characteristics. A logarithmic plot of the peroxide based sensor shows a linear response in the concentration range 5×10^{-8} to 5×10^{-4} M L-glutamate. The response levelled off at glutamate concentrations above 1 mM, and at a lower concentration of about 1.0×10^{-4} M with substrate recycling. The detection limits of a typical sensor are 30 nM and 90 nM for steady state and rate measurements, respectively. Although, the steady state measurement is more sensitive, the rate measurement has a larger linear range and a shorter response time of 14 sec compared to 2 min for steady state measurement. Figure 1 shows a comparison of the responses of the hydrogen peroxide based device with and without the 100 MW cut-off membrane and the oxygen based sensors with and without substrate recycling. The addition of a size-exclusion membrane reduced the sensitivity by about 50%, probably due to the slower diffusion of hydrogen peroxide through the less porous membrane. The sensitivity of the oxygen based device was also reduced to 60% by the addition of the hydrophobic oxygen membrane.

In the absence of NADPH, the responses of the monoenzyme and bienzyme electrodes were similar, indicating that co-immobilizing both enzymes does not cause any drawbacks. In the presence of 5 mM ammonium chloride and 1 mM NADPH, the response of the bienzyme system was amplified 10-fold. In the flow injection system, the rate response was found optimum at 0.80 ml/min while the peak current and assay time decreased non-linearly with the flow rate. The 0.80 mL/min was chosen as a compromise and fifty to sixty 100 μl samples/hr were assayed under these conditions. The linearity of a response curve obtained using an oxygen based electrode in conjunction with a flow system was similar to that obtained from a batch analysis.

L-Glutamate oxidase has been found to be very specific (13). However, a number of electroactive interferences from real samples can be expected at ∓ 650 mV. However, at -650 mV the oxygen based device with the hydrophobic oxygen membrane was found totally specific for L-glutamate except for samples which contain volatile electroactive species (e.g., volatile amines from hydrolyzed proteins). At +650 mV, a number of amino acids such as cysteine, tyrosine,

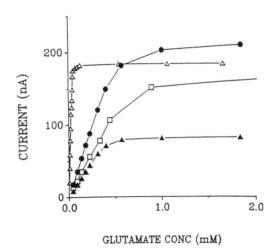

GLUTAMATE CONC (mM)

Figure 1.   Comparison of the steady-state current
response of the different Glutamate electrodes
constructed from Immobilon membrane. ● Hydrogen
peroxide based electrode using enzyme membrane only;
 ▲ hydrogen peroxide based electrode using enzyme
membrane with 100 MW cut-off membrane;   ☐ oxygen
based electrode using enzyme membrane and
hydrophobic oxygen membrane;   △ same as ☐ but with
substrate recycling.

dihydroxyphenylalanine and tryptophan and different neurotransmitters and their metabolites are electroactive and produce a false signal. However, the incorporation of the 100 MW cut-off membrane removed or at least reduced the interferences from these compounds to at most 3% when present in amounts equal to the L-glutamate concentration.

The proposed glutamate sensor was used to determine the glutamate content of 3 protein tablets and 11 food products. The results obtained for 2 of the 3 protein tablets and 8 of the 11 food products (Table III) correlated well with the manufacturer's specification with an acceptable range of error.

The positive interference may be caused by free glutamate or volatile amines from the protein hydrolyzate. The food products that did not contain hydrolyzed protein tested correctly. In addition, the low values obtained in the case A/G Tablets are due to the fact that some of the L-glutamate in these tablets may be bound as a peptide which the sensor could not detect.

Amperometric Aspartate Electrode. There is a growing interest in the rapid determination of aspartate because of its implications in neurochemistry and food analysis. However, our interest in an aspartate electrode stems from our interest in developing an aspartame electrode, since aspartate can be enzymatically cleaved from aspartame. The aspartate electrode is based on the co-immobilization of aspartate aminotransferase (AST) and glutamate oxidase (GMO). The reactions involved are:

$$\text{L-aspartate} + \alpha\text{-ketoglutarate} \xrightarrow{\text{AST}} \text{L-glutamate} + \text{oxaloacetate}$$

$$\text{L-glutamate} + O_2 + H_2O \xrightarrow{\text{GMO}} \alpha\text{-ketoglutarate} + NH_4^+ + H_2O_2$$

The hydrogen peroxide produced in the second reaction was detected at a platinum electrode held at $+650$ mV (versus Ag/AgCl) and the current change was linearly correlated to the concentration of L-aspartate present in solution. similar to the glutamate electrode, a size-exclusion membrane (100 MW cut-off) was inserted between the pt electrode and the enzyme membrane to eliminate interferences from electro-oxidizable species. Since the electrode can also detect glutamate, in samples containing both aspartate and glutamate, the signal from glutamate was eliminated by incorporating an outer membrane containing GMO and catalase.

Electrode Assembly. The electrode was assembled by placing on an inverted electrode jacket the following membrane:

(a) 1 cm^2 piece of the 100 MW cut-off cellulose acetate membrane;
(b) AST/GMO membrane prepared by crosslinking the two enzymes with BSA

Table III. Determination of Glutamate Content of Food Products and Protein Tablets

Product	Experimental Value[1] (%)	Manufacturer's Specification (%)	Relative Error (%)
Accent	95 ± 3	100	-5
McCormick Regular Season All	5.5 ± 0.1	5.8	-5.2
McCormick Spicy Season All	5.5 ± 0.1	5.8	-5.2
McCormick Seasoned Meat Tenderizer	10.0 ± 0.2	10.0	0
McCormick Unseasoned Meat Tenderizer	0	0	0
Adolph's Seasoned Meat Tenderizer	0	0	0
Morton Nature's Season	0	0	0
Mrs. Dash Extra-spicy Seasoning	0	0	0
Wylers Beef Bouillon Cube	5.2 ± 0.1	2	+160
Wylers Chicken Cube	4.8 ± 0.1	2	+140
Chun King Soy Sauce	15.8 ± 0.3	0	nd
Your Life Protein Tablet	8.4 ± 0.2	8.1	+3.7
Vitaline Free Form Amino Acids	2.2 ± 0.1	2.2	0
A/G Protein Tablet	0.75 ± 0.02	4.3	-83

[1]Assay values represent the average of 5 determinations with 95% confidence level

and glutaraldehyde and then covalently bonding on pre-activated Immobilon-AV affinity membrane; (c) 0.03 μm Nuclepore membrane with GMO/Catalase membrane. The membranes were then secured with an O-ring and parafilm. The electrode jacket was filled with PBS, pH 7.4 and the platinum/silver-silver chloride probe was then inserted into the jacket. A constant potential of +650 mV applied to the Pt working electrode. For measurements, the assembled electrode was equilibrated in PBS containing an appropriate amount of α-ketoglutarate.

Optimization of Parameters. It was mentioned earlier that the covalent immobilization of GMO on Immobilon-AV Affinity membranes gave the best compromise between activity and stability. However, the bienzymatic aspartate electrode assembled utilizing the same immobilization technique had a low current response, probably due to low enzyme loading since both enzymes compete for a limited number of binding sites on the membrane. To increase enzyme loading, the enzymes were mixed with a solution of BSA/glutaraldehyde then pipetted on the preactivated membrane. The optimum mixture contained 0.25 units GMO, 5.0 units AST, 0.05 mg BSA and 1 μl of 25% glutaraldehyde dissolved in 5 μl of 0.5 M KH_2PO_4, pH 7.4 on an Immobilon membrane. The GMO/Catalase membrane was prepared on a polycarbonate support using an immobilization matrix consisting of 5 μl of 400 units/ml GMO, 5 μl of 1000 units/ml Catalase, 10 μl of 10% BSA and 4 μl of 2.5% aqueous glutaraldehyde. The response of the electrode was optimum in the pH range 7 to 9. The observed optimum of the co-substrate, α-ketoglutarate is affected by both its concentration and the amounts of immobilized enzymes, since high concentrations of α-ketoglutarate shift not only the transminase reaction towards the production of glutamate, but also shift the GMO reaction towards the opposite direction. The optimum concentration shifted from 2.0 x 10^{-4} M to 1 x 10^{-3} M as the aspartate concentration was changed from 3.9 x 10^{-6} M to 1.4 x 10^{-3} M. As a compromise, 5 x 10^{-4} M α-ketoglutarate was used for further study.

Analytical Characteristics. Representative calibration curves were obtained in the presence and absence of different protective membranes. Without a second membrane, the response to aspartate is linear in the concentration range 1 x 10^{-6} M to 2.2 x 10^{-4} M and the response to glutamate is linear in the concentration range 8 x 10^{-7} M to 1.5 x 10^{-3} M. Addition of a 100 MW cut-off membrane and 0.03 μm polycarbonate membrane reduced the response to 50%. Incorporation of the GMO/Catalase membrane further reduced the response to 17%, but with a slight increase in the linearity range. Aspartate can be detected at 5 x 10^{-6} M with greater than 99.7% of the signal from glutamate in the sample being eliminated. The electrode is very stable when kept in Dulbecco's buffer at 4°C, retaining at lease 90% of its initial activity after 75 days. When stored dry, the sensitivity is reduced to less than 50% after 10 days. The electrodes proved to be very selective, and only asparagine and glutamine produced interferences of only 8% and 6%, respectively, when present in concentrations equal to aspartate. The determination of aspartate content of some pharmaceutical products and more information about the electrode were reported (14).

Amperometric Aspartame Electrode. Aspartame (N-L-α-aspartyl-L-phenylalanine methyl ester), is a low-calorie nutritive sweetener composed from two amino acids, L-aspartic acid and L-phenylalanine. The excellent selectivity of enzyme and microbial biosensors has led to the development of several sensors for the measurement of aspartame.   These include a microbial sensor(15) and potentiometric enzyme electrodes based on immobilized L-aspartase (16) and co-immobilized carboxypeptidase A and L-aspartase (17). However, these electrodes suffered from several interferences.   An assay technique using peptidase, aspartate aminotransferase and glutamate oxidase in solution was also reported (18).   However, analysis took at least 30 minutes and the enzymes were consumed. The proposed amperometric enzyme electrode is another application of the aspartate electrode previously described and based on the co-immobilization of the aspartame-hydrolyzing enzyme (APH), aspartate aminotransferase (AST) and glutamate oxidase (GMO). The reactions involved are:

$$\text{Aspartame + H}_2\text{O} \xrightarrow{\text{APH}}$$
$$\text{L-Aspartate + L-phenylalanine methylester}$$

$$\text{L-Aspartate + }\alpha\text{-Ketoglutarate} \xrightarrow{\text{ASF}}$$
$$\text{L-Glutamate + Oxaloacetate}$$

$$\text{L-Glutamate + O}_2\text{ + H}_2\text{O} \xrightarrow{\text{GMO}} \alpha\text{-Ketoglutarate + NH}_4^+ \text{ + H}_2\text{O}_2$$

Electrode Preparation and Measurement. The aspartame membrane was prepared by crosslinking APH, AST and GMO with BSA and glutaraldehyde, followed by covalently bonding on a pre-activated Immobilon-AV-Affinity membrane. 50 units of APH (BioEurope, Toulouse, France), 10 units AST, 1 unit GMO, 0.25 mg BSA and 2.5 μL of 0.25% glutaraldehyde were dissolved in 7.5 μl of 0.1 M KH$_2$PO$_4$, pH 7.0 and the mixture was immediately transfered onto the immobilon membrane (0.9 cm diameter).   The membrane was air dried for 2 hours and then kept immersed in buffer until ready to use. The electrode was assembled as described for the aspartate electrode. Solid food samples were first ground, then a weighed amount was dissolved in buffer. Liquid samples were tested without dilution. Prior to testing, air was bubbled through the samples for approx. 5 minutes. The electrode was immersed in a 10 ml beaker containing 5.00

ml of phosphate buffer with 5.0 x 10^{-4} M α-Ketoglutarate and stirred until the background current attained a stable value before sequentially injecting the aspartame samples.

Analytical Characteristics. The response of the electrode was linear in the concentration range 2.0 x 10^{-4} M to 1.5 x 10^{-3} M with a current slope of 2.1 nA/mM, correlation coefficient of 0.999 and a detection limit of 1.5 x 10^{-4} M, (S/N = 3). The response and equilibration times are 2-3 minutes, which is a significant improvement over the previously reported potentiometric enzyme electrodes.

The aspartame content of four food samples were determined using the proposed aspartame electrode. The results obtained correlated well with the manufacturer's specifications within 10% (Table IV).

Table IV. Determination of Aspartame in Food

Food Product	Aspartame Sensor	Manufacturer	Percent Error
Equal Tablet	22.5 ± 0.8 (wt %)	20.3 (wt %)	+ 10.8
Equal Powder	3.2 ± 0.2 (wt %)	3.5 (wt %)	- 8.6
Diet Coke	559 ± 33 mg/L	518 mg/L	+ 7.9
Diet Root Beer	504 ± 40 mg/L	490 mg/L	+ 2.9

Diet drinks exhibited some electroactive interferences, which were not eliminated by the size-exclusion membrane and could not be identified at this time. Consequently, the background current was first allowed to reach a steady-state, then α-ketoglutarate was added to initiate the transaminase reaction of aspartate. The current was measured after reaching another steady-state value, and the difference between the two steady-state currents was related linearly to the aspartame content of the sample.

Since aspartate aminotransferase and glutamate oxidase are both used in the immobilized matrix, the electrode is more sensitive to glutamate and aspartate than to aspartame. Thus, the assembled electrode is useful only for protein-free samples such as diet soft drinks and fruit juices.

Fiber Optic Sensors

The relatively new area of fiber optic based biosensors is one of the most active fields of chemical analysis today. The continued demand by the different regulatory agencies for new and improved methods for determining food quality and its suitability for consumption have stimulated a special interest in this area.

Fiber-optic biosensors are especially ideal for these systems that are not easily amenable to electrochemical analysis. Our current research focuses on the development of sensors for hydrogen peroxide, glucose, alcohol and sulfite in food products where some of the studies are still continuing.

Hydrogen Peroxide Fiber Optic Sensor. Fiber-optic sensors were developed for hydrogen peroxide determination based on absorbance and chemiluminescence measurements. In the spectrophotometric measurement, the use of the dye 2,2'-azinobis (3-ethylbenzthiazoline sulfonic acid) or ABTS in conjunction with the following enzymatic reaction, was found to be most promising

$$\text{Glucose} + O_2 \xrightarrow{\text{Glucose oxidase}} \text{gluconic acid} + H_2O_2$$

$$H_2O_2 + \text{ABTS(Colorless)} \xrightarrow{\text{POD}} \text{ABTS (green)} + 2H_2O$$

The decrease in signal due to the absorption of the produced colored dye at 425 nm was measured and related to the concentration of H_2O_2. The obtained calibration curves had linear dynamic ranges of 10^{-6} - 10^{-3} M and 10^{-5} - 10^{-2} M H_2O_2 using soluble and immobilized peroxidase, respectively. Good results were obtained for an assay of peroxide in tea. In addition, the micellar enhanced chemiluminescence of the peroxyoxalate and luminol/peroxidase systems were evaluated and exhibited linearity of 8×10^{-4} - 8×10^{-9} M and 1.2×10^{-4} to 2.4×10^{-8} M H_2O_2, respectively.

Glucose Fiber Optic Sensor. The developed hydrogen peroxide sensors were adapted for the measurement of glucose. The first glucose sensor was based on the spectrophotometric measurement of the color change of ABTS according to the following reactions:

$$H_2O_2 + \text{ABTS} \atop \text{(Colorless)} \xrightarrow[\text{Peroxidase}]{\text{Horseradish}} 2H_2O + \text{ABTS} \atop \text{(Green)}$$

Glucose oxidase was immobilized on Immunodyne membrane (Pall, NY), and mounted around the common end of a bifurcated fiber optic bundle. Monochromatic light at 425 nm was focused onto the input arm of the bundle and transported through the fiber to a light tight cell. The scattered radiation transported through the second arm was focused on a second monochromator, then measured with a photo-multiplier tube. The optimum conditions are: 0.075 M phosphate buffer, pH 7, 1 mM ABTS and 1 U of peroxidase. A linear relationship was observed between the attenuation of reflected light and glucose concentration in the range 1.1×10^{-5} to 1.5×10^{-3} M. The sensor was used for the determination of glucose in Gatorade Orange Drink and Pedialyte. Good results were obtained as compared to the AOAC and amperometric methods (Table V).

Table V.   Comparison Study of Glucose Methods

Method	Glucose (g/dl)	
	Pedialyte	Gatorade
AOAC	2.67 ∓ 0.07	3.01 ∓ 0.08
Amperometric	2.84 ∓ 0.04	3.02 ∓ 0.04
Fiber Optic/ABTS	2.73 ∓ 0.05	3.30 ∓ 0.07
Fiber Optic/CL		3.16 ∓ 0.11

The oxidized ABTS is irreversibly adsorbed on the membrane surface with continuous use, which may be the major cause for the decrease in the enzyme activity with time.

The determination of glucose using micellar enhanced chemiluminescence measurement of the luminol and peroxyoxalate reactions was also investigated (19,20). Although good linearities of 3×10^{-7} - 3×10^{-4} and 3×10^{-6} x 10^{-2} M glucose were obtained, only the peroxyoxalate system was utilized for the assay of glucose in Gatorade (Table V). The preliminary results suggest that the glucose analysis is feasible using normal micelles at an acceptable pH where the activity of immobilized enzyme is not inhibited.

Sulfite Fiber Optic Sensor. Sulfite is extensively used as a preservative in food products. As a result of the increasing concern about the potential health hazards, the Food and Drug Administration (FDA) has recommended a 10 ppm sulfite limit in food. Preliminary studies to determine sulfite using immobilized sulfite oxidase coupled to the previously described colorimetric and chemiluminescent hydrogen peroxide sensors were partially successful, and further improvements are still possible. Additionally, we have shifted attention to the development of a sulfite sensor based on an enzymatic/oxygen fluorescence quenching approach. An oxygen sensitive membrane was prepared from a mixture of perylene dissolved in siloprene crosslinking agent and siloprene solutions. The membrane was mounted on the distal end of the bifurcated fiber optic bundle, covered by an enzyme membrane of sulfite oxidase immobilized on an immunodyne membrane, then secured in place with a dialysis membrane and an O-ring. The oxidation of sulfite by sulfite oxidase consumes oxygen and decreases its concentration in the solution, reducing the efficiency of oxygen quenching. As a result, the fluorescence intensity of the indicator increases proportional to sulfite concentration. Typical calibration curves were linear in the concentration range 2-100 ppm sulfite with a response time of 5-25 minutes depending on sulfite concentration. The standard deviation of five successive analysis of each of 10 and 30 ppm sulfite standard solutions were 0.24 and 1.15, respectively. The sulfite content of a limited number of food samples (dry apricots, raisins, potato chips and lemon juice) were determined with the proposed sensor and the results obtained correlated well with those obtained with the AOAC method. Although the final optimization and evaluation of the proposed sensor has not been completed, the methodology seems very promising.

Acknowledgments

The financial support of the US Department of Agriculture in the form of SBIR Grants (86-SBIR-8-0096) and (89-33610-4318) is gratefully acknowledged. The authors wish also to acknowledge the commendable contribution of their research group, especially Drs. R. Villarta, X. Xie and Z. Shakhsher.

Literature Cited

1. Henniger, G.; Mascaro, L., Jr. J. Assoc. Off. Anal. Chem., 1985, 68, 1021.
2. Guilbault, G. G.; Handbook of Enzymatic Methods of Analysis; Marcel Dekker, New York, NY, 1976;
3. Schubert, F.; Kirstein, D.; Scheller, F. Anal. Lett., 1986, 19, 2155.
4. Ameyama, M. Methods in Enzymology, 1982, 89, 20.
5. Nakashima, K.; Takei, H.; Adachi, O.; Shinagawa, E.; Ameyama, M. Clin. Chim. Acta, 1985, 151, 307.
6. Matsumoto, K.; Hamada, O.; Ukeda, H.; Osajima, Y. Anal. Chem., 1986, 58, 2732.
7. Matsumoto, K.; Kamikado, H.; Matsubara, H.; Osajima, Y. Anal. Chem., 1988, 60, 147.
8. Xie, X.; Kuan, S. S.; Guilbault, G. G., Biosensors & Bioelectronics, 1991, 6, 49.
9. Schaumburg, J. J.; Byck, R.; Gerstl, R.; Mashman, J. H., Science, 1969, 163, 826.
10. Yamanchi, H.; Kusakabe, H.; Midorikawa, Y.; Fujishima, T.; Kuninak, A., Eur. Congr. Biotechnol., 1984, 1, 705.
11. Villarta, R. L.; Cunningham, D. D.; Guilbault, G. G., Talanta, 1991, 38(1), 49.
12. Villarta, R. L. Ph.D. Thesis, University of New Orleans, New Orleans, LA (1991).
13. Kusakabe, H.; Midorikawa, Y.; Fujishima, T.; Kuninaka, A.; Yoshino, H., Agri. Biol. Chem., 1983, 47, 1323.
14. Villarta, R. L.; Palleschi, G.; Lubrano, G. J.; Suleiman, A. A.; Guilbault, G. G., Anal. Chim. Acta, 1991, 245, 63.
15. Renneberg, R.; Riedel, K.; Scheller, F., Appl. Microbial. Biotechnol., 1985, 21, 180.

16. Guilbault, G. G.; Lubrano, G. J.; Kauffmann, J-M.; Patriarche, G. J., Anal. Chim. Acta, 1988, 206, 369.
17. Fatibello-Filho, O.; Suleiman, A. A.; Guilbault, G. G.; Lubrano, G. J., Anal. Chem., 1988, 60, 2397.
18. Mulchandani, A.; Male, K. B.; Luong, J. H. T.; Gibbs, B. F., Anal. Chim. Acta, 1990, 234, 465.
19. Abdel-Latif, M. S.; Guilbault, G. G., Anal. Chim. Acta, 1989, 221, 11.
20. Abdel-Latif, M. S.; Guilbault, G. G., Anal. Chem., 1988, 60, 2671.

RECEIVED July 13, 1992

Chapter 4

Metalloporphyrin-Coated Electrodes for Detection of 2,4-D

Namal Priyantha and Marianne Tambalo

Department of Chemistry, University of Hawaii at Hilo, Hilo, HI 96720

Glassy carbon electrodes coated with 5,10,15,20-tetraphenyl porphyrinato cobalt(II) [Co(II)TPP] and 5,10,15,20-tetraphenyl porphyrinato iron(III) chloride [Fe(III)TPPCl] provide an electrocatalytic amperometric sensor for detection of the herbicide, 2,4-D, and related compounds. The linear dynamic range of the sensor for 2,4-D is 2×10^{-6} M to 1×10^{-5} M with the lower detection limit of 1×10^{-7} M.

The compound, 2,4-dichlorophenoxyacetic acid (2,4-D) and related compounds such as 2,4,5-trichlorophenoxyacetic acid (2,4,5-T), 2-(2,4,5-trichlorophenoxy)propionic acid, 4-chloro-2-methylphenoxyacetic acid (MCPA) and 2,4-dichlorophenoxyethyl sulfuric acid have outstanding and unusual herbicidal properties. The herbicide 2,4-D is highly toxic to most broadleaf plants and relatively nontoxic to monocotyledonal plants. Consequently, it is used for post-emergence control of annual and perennial broadleaf weeds in cereals, sorghum, sugarcane, and on non-crop land, including areas adjacent to water (1,2). There are several metabolic pathways for 2,4-D in the environment such as hydroxylation, decarboxylation, cleavage of the acid side-chain and ring opening.

Colorimetric and gas chromatographic methods with a variety of detectors have been used for analysis and detection of 2,4-D in environmental and agricultural samples (3,4). Both methods require derivatization of the 2,4-D

0097–6156/92/0511–0041$06.00/0

molecule either to make it more light sensitive or more volatile. Derivatization is usually cumbersome, expensive and occasionally hazardous. Thus, there is a great necessity for development of inexpensive and non-toxic detection techniques for herbicides.

Metalloporphyrins and related cyclic compounds are known to catalyze a variety of reactions including the reduction of oxygen (5,6) and organic compounds with halogen substituents (7,8). Most of these porphyrin catalysts have been used in homogeneous solutions of nonaqueous solvents, although there have been few reports in aqueous media. The reduction of organohalides occurs via nucleophilic aromatic substitution, and the formation of a porphyrin-metal-aryl adduct intermediate has been demonstrated recently (9). Reductive cleavage of this intermediate occurs during the voltammetric potential scan yielding several products and the starting form of the catalyst (10).

Electrocatalysis can be employed to overcome sluggish electron transfer kinetics of many analytes. The use of monomeric porphyrin coatings on electrode surfaces to demonstrate analytical applications of organohalides through electrocatalysis was reported recently (11). Polymerized porphyrin coatings on electrode surfaces have also been successfully characterized using electrochemical techniques (12, 13). However, the application of porphyrin modified electrodes for detection of pesticides has been limited, although a variety of electroanalytical techniques have been successfully introduced in this area with the aid of other types of catalysts (14, 15).

In our studies, tetraphenylporphyrinato cobalt (II) [Co(II)TPP] and tetraphenyl porphyrinato iron (III) chloride [Fe(III)TPPCl] coatings on glassy carbon electrode surfaces are used as electrocatalysts for the reduction of 2,4-D and related compounds. The applicability of the porphyrin coated electrodes operated at low potentials was demonstrated for amperometric detection.

EXPERIMENTAL

Materials

5,10,15,20,-Tetraphenyl porphyrinato cobalt (II) [Co(II)TPP] and 5,10,15,20-tetraphenyl porphyrinato iron (III) chloride [Fe(III)TPPCl] were purchased from Aldrich Chemical Co. and used as received. 2,4-Dichloropheonoxyacetic acid (2,4-D), 2,3-dichlorophenoxyacetic acid (2,3-D), and 4-chlorophenoxyacetic acid (4-CPA) were purchased from Aldrich Chemical Co. and purified by solvent extraction for electrochemical experiments. Dichloromethane and acetonitrile solvents were analytical grade and distilled prior to use. All supporting electrolytes were analytical grade, and aqueous electrolyte solutions were prepared from freshly distilled deionized water. Deaeration was accomplished by sparging with purified nitrogen, and all experiments were carried out under an N_2 atmosphere. The solvent system for all experiments was 25% CH_3CN and 75% water. The substrates injected for amperometric experiments were completely degassed with purified N_2 prior to injection to remove dissolved oxygen.

Instrumentation

Cyclic voltammetry was performed with an EC/225 voltammetric analyzer (IBM Instruments) and recorded on a Houston Instruments model 200 X-Y recorder. Amperometric experiments were conducted with a CV-1B cyclic voltammograph (Bioanalytical Systems Inc.) and recorded on a Linear Instruments model 1200 strip chart recorder. All potentials are reported with respect to Ag/AgCl reference electrode. Glassy carbon (GC), platinum wire and Ag/AgCl electrodes were used as working, counter and reference, respectively, for all voltammetric and amperometric experiments. Glassy carbon electrodes were coated with the porphyrin catalyst as described in an earlier publication (11).

RESULTS AND DISCUSSION

Aromatic carboxylic acids with chloro substituent groups, such as 2,4-D, 2,3-D, and 4-CPA are not easily reducible at bare electrodes. Cyclic voltammetric experiments of these substrates conducted at bare GC electrodes in 0.1 M chloride electrolytes showed that the reduction does not occur until -1.3 V, which is very close to the hydrogen evolution potential. The reduction observed is probably due to the direct reduction of the carboxylic acid group in the molecules. However, this potential is not suitable to design a detection scheme due to high background current and possible electrochemical interference.

It has already been conclusively demonstrated that the metalloporphyrin films catalyze the reduction of aromatic and aliphatic organohalides via a "chemical electrocatalysis" mechanism (8,16). The first reduction of 2,4-D at Co(II)TPP and Fe(III)TPPCl coated electrodes occurs at low potentials (figure 1), and the detailed electrocatalytic mechanism has been reported in a previous publication (11). Hence, the constant applied potential of -0.20 V was selected for all steady state amperometric experiments. At this potential, the background current and the effect of interference can be minimized.

According to the suggested mechanism, the electrocatalytic reduction current generated at the electrode surface is directly proportional to the bulk concentration of the analyte, which is the basis of the proposed amperometric sensor. The Co(II)TPP based sensor produced a stable baseline after about two minutes allowing for the background current to decay. The current at -0.20 V constant applied potential was increased with injections of 2,4-D as expected from electrocatalytic properties of the metalloporphyrin coatings.

The injections were continued until the saturation limit was reached (figure 2). The amperometric response of bare GC electrode at the same potential was significantly smaller compared to that at the coated electrode.   The responses at the coated electrodes were used to produce a calibration curve for 2,4-D which shows two linear dynamic ranges over the concentration range of 2×10^{-6} to 1×10^{-5} M (figure 3).   The presence of two linear portions has been observed with systems where electrocatalysis takes place due to mediated electron transfer (17). However, the exact reason is still unknown.

The lower detection limit of the Co(II)TPP sensor was estimated to be 1×10^{-7} M based on the signal-to-noise ratio of 2.   This detection mechanism was very promising due to the fact that 2,4-D is not able to be detected at a potential below -1.3 V with a bare electrode. The linear detection range can be shifted to higher concentrations when a larger amount of analyte is injected instead of 5 uM steps.

Several other metalloporphyrins, electrolytes, and analytes   were investigated in order to demonstrate the applicability of metalloporphyrin electrode for the detection of chloro-organic   compounds.   Although Fe(III)TPPCl coated electrode is less sensitive than Co(II)TPP electrode, all the three analytes studied, 2,4-D, 2,3-D and 4-CPA, showed similar linear dynamic ranges (figure 3,4) with a lower detection limit of the same order of magnitude (table I) at both types of electrodes.

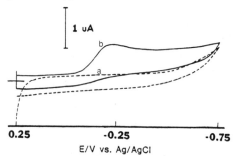

1 uA

b

a

0.25   -0.25   -0.75

E/V vs. Ag/AgCl

NOTE: Voltammograms were labeled as a and b.

Figure 1.   Cyclic voltammograms of 5×10^{-3} M 2,4-D at
                       (a) bare, (b) Co(II)TPP coated GC electrode in 0.1 M
                       LiCl under N_2 satd. Scan rate 10 mV/sec.

CONCLUSION

The catalytic activity of Co(II)TPP and Fe(III)TPPCl coated glassy carbon electrodes towards the reduction of 2,4-D, 2,3-D,   and 4-CPA enables the amperometric detection of these substrates at a relatively low potential. This catalytic scheme demonstrates the potential applicability of metalloporphyrin coated electrodes as amperometric detectors for real samples,   containing herbicides with chlorine substitued phenoxy acetic   acids, after liquid chromatographic separation (LCEC).

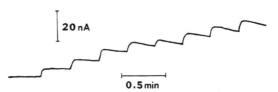

Figure 2.   Steady state amperometric (current vs. time) responses
                     obtained with the Co(II)TPP coated GCE with
                     increasing concentrations of 2,4-D in 5 uM steps (10 uL
                     additions) under N_2 satd. Applied potential -0.20 V vs.
                     Ag/AgCl, supporting electrolyte 0.1 M LiCl in
                     H_2O/CH_3CN (3:1).

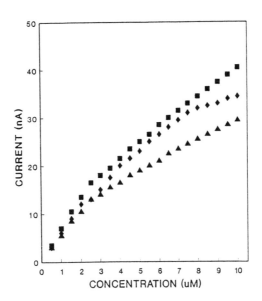

Figure 3.   Calibration curves at Co(II)TPP coated GC electrode
                     (♦) 2,4-D (▲) 2,3-D (■) 4-CPA. Supporting
                     electrolyte 0.1 M LiCl in H_2O/CH_3CN (3:1).

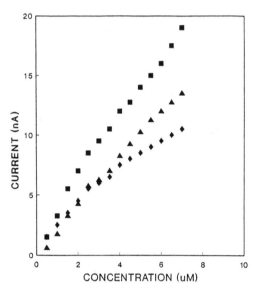

Figure 4.   Calibration curves at Fe(III)TPPCl coated GC electrode (♦) 2,4-D (▲) 2,3-D (■) 4-CPA. Supporting electrolyte 0.1 M LiCl in H_2O/CH_3CN (3:1).

Table I: Sensitivity and detection limits of amperometric detection

Compound	Sensitivity (nA mol^{-1} L)		Det. Lim.
	Co (II)TPP	Fe(III)TPPC1	
2,4-D	3.2×10^6	1.2×10^6	1×10^{-7}
2,3-D	2.2×10^6	1.8×10^6	1×10^{-7}
4-CPA	3.2×10^6	2.2×10^6	1×10^{-7}

ACKNOWLEDGEMENT

The authors greatly appreciate the financial support from the Office of Research Administration, University of Hawaii (Grant No: R-92-867-F-728-B-279).

REFERENCES

1.   The Royal Society of Chemistry, Agrochemicals Handbook, Second edition, Unwin Brothers: Old Woking, Surrey, 1987.
2.   Hagin, R. D.; Linscott, D. L.; Dawson, J. E. J. Agric. Food Chem. 1970, 18, 848.
3.   McCall, P. J.; Vrona, S. A.; Kelley, S.S. J. Agric. Food Chem. 1981, 29, 100.
4.   World Health Organization, Environmental Health Criteria 29, Geneva, 1984, p. 17.
5.   Kazarinov,V. E.; Tarasevich, M. R.;Radyushkina, K.A.; Andreev, V. N. J. Electroanal. Chem. 1979, 100, 225.
6.   Ouyang, J.; Anson, F. C. J. Electroanal. Chem. 1989, 271, 331.
7.   Lexa, D.; Saveant, J. M. J. Am. Chem. Soc. 1982, 104, 3503.
8.   Lexa, D.; Saveant, J. M.; Su, J. B.; Wang, D. L. J. Am. Chem. Soc. 1987, 109, 6464.
9.   Kadish, K. M. Prog. Inorg. Chem. 1986, 34, 435.
10. Kadish, K. M.; Lin, X. Q.; Han, B. C. Inorg. Chem. 1987, 26, 4161.
11. Root, D. P.; Pitz, G.; Priyantha, N. Electrochim. Acta 1991, 36, 855.
12. White, B. A.; Murray, R. W. J. Electroanal. Chem. 1985, 189, 345.
13. Deronzier, A.; Latour, J. J. Electroanal. Chem. 1987, 224, 295.
14. Mendez, J. H.; Martinez, R. C.; Gonzalo, E. R.; Trancon, J. P. Electroanal. 1990, 2, 487.
15. Pingarron, J. M.; Gonzalez, A.; Polo, L. M. Electroanal. 1990, 2, 493.
16. Elliott, C. M.; Marrese, C. A. J. Electroanal. Chem. 1981, 119, 395.
17. Wang, J.; Lin, M.S. Electroanal. 1989, 1, 43.

RECEIVED June 8, 1992

Chapter 5

Intact Chemoreceptor-Based Biosensors

Antennular Receptrodes

R. Michael Buch

Warner-Lambert Company, 170 Tabor Road, Morris Plains, NJ 07950

As evidenced by the collection of works assembled in this compilation, a tremendous amount of innovative experimentation is being applied to the field of biosensor research. Although many different biosensors have been developed during the past decade, all share two common components: a molecular recognition element and a transducer. The molecular recognition element is the biological component of the biosensor that provides the device with a degree of chemical selectivity. The transducer, on the other hand, is the nonbiological portion of the biosensor which converts the chemically-coded information received at the molecular recognition element into a measurable signal (usually electrical). This chapter will describe a novel type of biosensor, termed an antennular receptrode, in which both the molecular recognition element and the transducer are biological entities.

The unique binding properties of chemoreceptors make them ideally suited to function as molecular recognition elements in biosensors. Many chemoreceptors are extremely specific, serving as binding sites for only one type of biological messenger. However, difficulties associated with isolation, stabilization, and immobilization of these complex macromolecules arise when attempts are made to incorporate them into biosensors. Our approach has been to circumvent these difficulties by utilizing an intact chemosensing organ, which permits the chemoreceptor molecules to remain within their native (optimal) biological environments, the chemoreceptive neurons.

The chemosensing structures (antennules) of several species of crustacea have been employed in this research. Earlier work utilized the intact antennules of *Callinectes sapidus*, the Atlantic blue crab.(1,2) These earlier

0097–6156/92/0511–0048$06.00/0

receptor-based biosensors were able to selectively detect the 21 naturally-occurring amino acids at the nanomolar level. Further experimentation resulted in the detection of several excitatory amino acids below the picomolar level. More current work, to be discussed in the remainder of this chapter, focusses on the development of neuronal-based biosensors utilizing the antennules of two Pacific crustaceans, *Podophthalmus vigil*, the Hawaiian long-eyed crab, and *Portunis sanguinolentus*, the Hawaiian blood-spotted crab. The incorporation of chemoreceptors from different species has extended the number and types of compounds detected by this new class of biosensor, while simultaneously demonstrating that the construction of antennular receptrodes is not species limited. Furthermore, the small size of the Pacific species' antennular neurons (relative to *C. sapidus*) precipitated the development of a new digital data sampling technique and new data analysis algorithms. Before any discussion of neuronal-based biosensors can begin, a brief description of antennular chemoreception is necessary.

Physiology of antennular chemoreception

Because eyesight is of little use in the turbid waters typically inhabited by crabs, they have developed very sensitive chemical senses. Crabs use these senses (olfaction and gustation) to locate food and mates, to recognize territory, and to detect potential danger. The principle chemosensory organs of crabs are the antennules, the periopod dactyls, and the mouthparts. Because gustation and olfaction occur in the same aqueous environment for aquatic species, an arbitrary parameter of distance is employed to distinguish between the two senses. A chemosensing organ responding to chemical stimuli originating from distant sources is termed an olfactory organ. Similarly, an organ responding to stimuli from a local source is termed a gustatory organ. Because antennules are olfactory organs, they are inherently more sensitive than gustatory organs. The ability to function with this high degree of sensitivity in aqueous solution was the major reason why the antennules were the organ of choice for this research. A secondary consideration was the fact that the intact antennules could be easily excised from the organisms.

P. sanguinolentus is physically very similar to *C. sapidus*, having two antennules, each measuring ≈1cm in length and ≈1mm in diameter located between the eyestalks on the anterior portion of the carapace. *P. vigil* is a smaller crab with disproportionately long periopod dactyls (claws), eyestalks, and antennules. The antennules, measuring approximately 2cm, are similar to those of the other two species in all aspects but length. The antennules of each of these species are jointed. The first joint connects the coxa segment to the carapace. The coxa segment is connected to the basis segment by a second elbow-like joint. The tip of the basis segment (the distal tip of the antennule) is a biramous structure. The larger of the two branches is called the endopod and the smaller branch is called the exopod.

The endopod contains a tuft of several hundred hair-like sensillae, known as aesthetascs. Electron microscopic studies reveal the aesthetascs to

be open-ended thin-walled tubules protruding from the flagella.(3)   The
aesthetascs contain the dendrites of many sensory neurons (chemoreceptive
cells) which contain an even greater number of chemoreceptors.(4)
Chemoreceptors are actually protein molecules embedded in and across the
dendritic membranes. These transmembrane protein molecules contain one
or more specific binding sites.   In the case of olfactory receptors, the
compounds binding at these specific sites are called odorants.   Thus the
olfactory chemoreceptors function as target molecules for specific odorants.
Dye penetration experiments indicate that the aesthetascs are permeable to
liquids, thereby allowing odorants to directly contact the chemoreceptors.(5)

A chemoreceptive neuron is said to be in the "resting state" when no
odorants are present.   During this resting state, the concentration of sodium
ions is ≈ 10 times greater on the exterior of the cell than in the interior
(axoplasm).   Conversely, the concentration of potassium ions is ≈ 40 times
greater internally than externally. This non-equilibrium condition, maintained
by the well-documented ATP/ATP-ase driven sodium pump mechanism, results
in a transmembrane potential of $\approx -90mV$.   Although potassium ions can
traverse the cellular membrane freely, the internal electronegativity resulting
from the efflux of sodium ions causes the potassium ions to concentrate within
the axoplasm.(6)

Although no definitive structural studies have been performed on
crustacean olfactory receptors, evidence indicates that these receptors function
similarly to the more thoroughly studied nicotinic acetylcholine receptor. The
acetylcholine receptor is composed of 5 sub-units assembled in a manner which
allows a channel to traverse the entire length of the molecule. Because the
molecule is a transmembrane protein, the channel is a transmembrane channel.
Several specific binding sites are located on the portion of the molecule which
extends into the cellular exterior.   In the resting state, when none of the
binding sites are occupied, the channel is closed. When the binding sites are
occupied (in the presence of specific odorants), the tertiary structure of the
chemoreceptor changes, effectively opening the channel. Ions can now follow
the potential gradient by migrating through the channel.   A consequence
of the influx of sodium ions is a change in the membrane potential. This small
localized change triggers a cascade of events (via voltage-gated ion channels)
which results in a depolarization of the local region of the neuromembrane.
The ionic flux does not bring the membrane potential immediately to ground,
but actually overshoots, making the exterior of the membrane negative with
respect to the interior. This negative afterpotential is followed by an overshoot
in the opposite direction, creating a positive afterpotential. Because the initial
odorant-receptor binding event is transitory, after several milliseconds the local
membrane potential is restored via the sodium pump mechanism. The voltage
pulse, in conjunction with the negative and positive afterpotentials, is termed
an "action potential".

The depolarization of one membrane region triggers voltage-gated
receptors in the neighboring regions of the neuromembrane. In this manner,
the depolarization is propagated along the entire length of the neuron. The

action potential is propagated along the length of the neuron until it reaches the terminus, known as the presynaptic fiber, named for the very small gap ($\approx 1\mu m$), called the synapse, between adjacent neurons.

The depolarization of the presynaptic fiber initiates an exocitotic process by which messenger compounds (e.g., acetylcholine) are released. The messenger compounds migrate across the synapse to the post synaptic fiber of the adjacent neuron, where they bind to specific receptors. These binding events then trigger the propagation of the action potential along the length of the neighboring neuron. Because this propagation is an active process, driven by cellular metabolism, no attenuation of the signal occurs. The entire action potential process occurs in approximately two milliseconds.

Because the number of action potential spikes produced (frequency) is directly proportional to the number of receptor-agonist binding events, and the number of binding events is proportional to the concentration of agonist (analyte), the frequency of the action potential can be correlated to the concentration of a particular analyte. The relationship between a neuron's response (R) and the concentration of agonist (C) is given by the following equation:

$$R = R_{max}/[1+(K/C)^n]$$

where R_{max} is the maximum response frequency, n is the Hill coefficient (a cooperativity factor between receptors), and K is a constant. For the most simplistic case, assuming identical receptors where $n=1$, the equation reduces to the more familiar Michaelis-Menton equation for enzyme kinetics.

From this rather rudimentary explanation of antennular chemoreception, it is obvious that a biosensor based on the concentration-frequency relationship could be created if the action potential could be "intercepted" during propagation. The apparatus and methodology used to accomplish this task are described below.

The Neuronal Biosensor

A plexiglass flow cell, consisting of several chambers, was used to hold the dissected antennule in place. The undissected chemosensing tip (endopod and exopod) of the antennule was inserted into a tubular chamber through which a flowing stream of artificial sea water (ASW), prepared according to the Woods Hole formula,(7) was pumped. Adjacent to this chamber and connected by a small mounting hole, was another chamber in which the dissected portion of the antennule (exposed antennular nerve) was bathed in an artificial isotonic intercellular fluid known as Panulirus saline solution.(8) Ground and reference wires were inserted into this chamber. Surrounding these chambers was a larger chamber through which thermostated water was pumped to maintain a constant temperature of 22°C, the normal environmental temperature of the Hawaiian Crabs. The entire flow cell was mounted on the stage of binocular dissecting microscope. A peristaltic pump was used to pump the carrier stream (ASW) over the chemosensing tip. Stimulant (analyte) solutions were introduced into the carrier stream by a four-way flow injection valve equipped with 0.5 ml sample loop.

Action potentials were "intercepted" with a glass "pick-up" electrode constructed from a $50\mu l$ capillary pulled to $20\mu m$ I.D. tip. This capillary was mounted on a microelectrode holder fitted with a vacuum line connected to a 1ml tuberculin syringe. The electrode holder was mounted on a mechanical micromanipulator. Electrical contact was established through a silver wire, mounted inside the electrode holder. This wire extended to within 2mm of the electrode tip. The lead from the wire was fed into a physiological preamplifier, the outputs of which were fed into an audio speaker and a computer equipped with an oscilloscope emulator. More detailed descriptions of the experimental apparatus can be found in several publications.[1,2]

Methodology

Identical procedures were employed for both Hawaiian crab species. A specimen was chilled for approximately 2min, the time required to completely immobilize these poikilothermic organisms. Dissecting scissors were then used to sever the coxa segment of an antennule (which could be regenerated by the animal within several weeks) as close to the carapace as possible. The excised antennule was transferred to a microscope slide and submerged in a drop of Panulirus saline solution to prevent desiccation. The slide was then placed on the stage of a dissecting microscope, where microdissection was performed to remove the exoskeleton of the coxa segment. The underlying connective tissue was removed to provide access to the antennular nerve, a large (0.5mm diameter) bundle of neurons. The exposed nerve was then dipped in a micromolar solution of trypsin for approximately 4s to enzymatically remove the thin layer of glial cells surrounding the nerve. The removal of these insulating cells provided a more favorable signal to noise ratio by allowing better contact between the pick-up electrode and the neurons.

The dissected antennule was mounted in the flow cell by inserting the biramous tip through the small (1mm diameter) mounting hole and into the flow chamber, through which the ASW carrier stream was pumped. Any gaps between the antennule and the walls of the mounting hole were filled with dental wax, thereby isolating the flow chamber from the neurobathing chamber, in which the exposed neurons were submerged in Panulirus saline solution. After connecting the various tubing lines (ASW flow and thermostated water), the ground and reference wires as well as the pick-up electrode, were inserted into the neurobathing chamber. Application of mild suction via the tuberculin syringe allowed the Panulirus saline solution to fill the electrode and complete the electrical circuit. After switching on the various electronic components, the micromanipulator was used to position the pick-up electrode adjacent to the antennular nerve. As the suction was increased, the nerve tissue was drawn against the tip of the electrode, and action potentials appeared on the computer monitor. The action potentials could also be heard as a "crackling" noise over the audio speaker, enabling the researcher to know when the proper electrical contact had been established without looking away from the microscope.

After electrical contact had been established, various analytes were introduced into the carrier stream. If no increase in action potential frequency was observed, the electrode would be repositioned and the process repeated. When a noticeable change in frequency occurred, varying concentrations of the particular analyte were introduced into the carrier stream. These solutions were introduced in order of increasing concentration to minimize adaptation effects caused by receptor binding site saturation. Computer programs written in C programming language sorted the action potentials according to amplitude, then generated a histogram (and table) of the number of spikes occurring within each amplitude range per given period of time. This data sorting method proved to be necessary due to the complex nature of the multi-unit data generated by the biosensor. By examining the frequency changes elicited by different concentrations of analyte within each amplitude range, the amplitude of interest, i.e. the amplitude with a quantitative frequency/dose response could be selected. The frequencies of the amplitude of interest corresponding to the neuron or population of neurons of interest were then plotted against the concentration of analyte to produce a dose/response curve, which was essentially a calibration curve for the biosensor.

Typical Data

Two types of data produced by antennular receptrodes have been reported previously.(9) These two types of data are termed single-unit data, which consist of action potential spikes of identical amplitudes, and multi-unit data, which consist of spikes of varying amplitudes. Single unit data are easily analyzed because the spikes can simply be counted for a given period of time to produce a resultant frequency. Single unit spikes are observed when the electrode is in contact with only one neuron or a population of neurons with similar properties (i.e. receptors).

A difficulty associated with the use of Hawaiian species is the fact that only multi-unit responses have been observed. This type of response, depicted in Figure 1, is more difficult to interpret because the action potential generated by the chemosensing neuron of interest is often masked by other spikes; thus, a simple count of the spikes is quantitatively meaningless. The fact that different amplitudes are observed is an artifact of the electrode/nerve fiber contact (all action potentials are approximately 90mV as determined by the transmembrane potential). For example, direct contact between the electrode and neuron results in a comparatively large amplitude. If a neuron is contacted poorly, the amplitude will appear to be relatively small. Thus, a multi-unit response is observed when an electrode is in contact (to varying

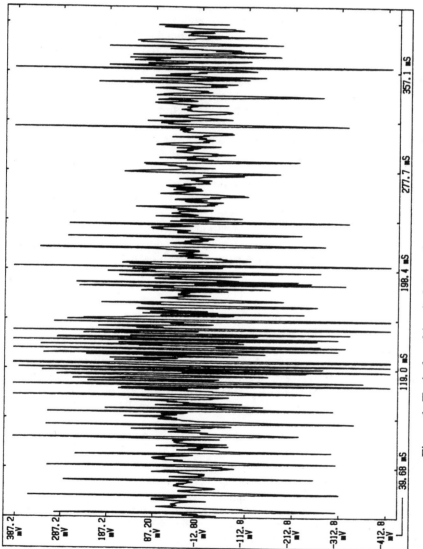

Figure 1: Typical multi-unit data as they appear on computer monitor.

extents) with several neurons. Because of the small size of the Hawaiian species' antennular neurons relative to the electrode tip, single unit responses were not observed. Single unit responses could presumably be achieved with a smaller electrode tip, capable of contacting only one neuron; however, a smaller tip creates a much higher impedance, which, in turn, makes the system more susceptible to noise. For this reason, the computerized data sorting technique was the method of choice.

The utility of the computerized data sorting technique is illustrated by the system's response to TMO, trimethylamine oxide. TMO is a degradation product of nitrogenous plant and animal substances, and, as a potential food marker, would be expected to elicit a response in an organism. Lockwood(10) previously reported *P. sanguinolentus* responses to relatively high concentrations (0.01M to 0.1M) of TMO. Because of this previous work, TMO was employed as a model analyte for our antennular receptrode studies.

The antennular receptrodes proved to be extremely sensitive to TMO. Although true concentrations cannot be accurately calculated at the extremely low levels found to evoke responses from the system, the concentrations were certainly well below the picomolar level. Concentrations of TMO calculated to be ca. 10^{-14}M produced easily measurable signals. At these low concentrations, loss of analyte due to adsorption onto the walls of the containers and tubing lines becomes significant, resulting in concentrations lower than those calculated. The result is that the system responds to concentrations even lower than those calculated. Figure 2 depicts typical histograms displayed by the computer following two injections of different concentrations of TMO. The concentrations in this figure were calculated (based upon serial dilution) to be $\approx 10^{-14}$M and 10^{-15}M. By comparing the two histograms, one can see that the amplitudes in the third "bin" undergo the largest change versus concentration. This amplitude bin corresponds to action potentials with frequencies between 201 and 256mV. When these frequencies are plotted against analyte concentration, a dose/response curve for the receptrode can be generated, as shown in Figure 3. This curve illustrates the low detection limit and broad dynamic range achieved with the antennular receptrode. Similar responses have been generated with various analytes, including glutamate, alanine, kainic acid, taurine, ADP, and AMP. Quantitative results have been obtained with both *P. sanguinolentus* and *P. vigil*.

Response times of antennular receptrodes are on the order of milliseconds. The mechanism responsible for the extremely rapid response has been described by Thompson and Krull.(11) They theorize that the mechanism is rapid because equilibrium or steady-state conditions must not be attained. Instead, transient perturbations alter a pre-existing non-equilibrium condition (the neuron's resting state), producing a large signal before the system can establish a new equilibrium or steady-state condition energetically suitable for the altered system. Because no new equilibrium or steady-state condition exists, very little time is required to regenerate the original (resting state) conditions.

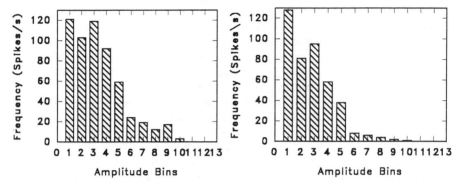

Figure 2: Frequency sorting histograms and the corresponding data tables generated by an antennular receptrode constructed from the antennule of *P. vigil* upon exposure to ≈10-14M (left) and ≈10-15M (right) TMO. Note the significant increase in the amplitude bin 3, corresponding to amplitudes between 201mV and 256mV.

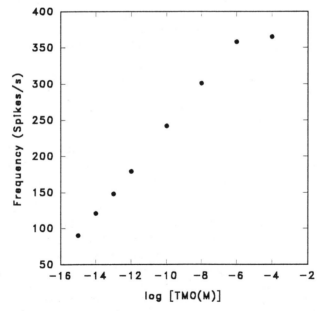

Figure 3: A typical dose/response curve for an antennular receptrode constructed from the intact chemosensing structures of *P. vigil*.

Although the response times are extremely rapid, throughput is limited by another phenomenon, known as adaptation, which can be described as receptor saturation.   If new stimuli (analytes) are introduced before the previous stimulant molecules have dissociated from the receptor binding sites, the new stimulant molecules will be sterically prevented from binding, and, therefore, will not be detected by the receptrode system.   The limit to throughput is the time required for a receptor-analyte complex to dissociate. Three minute exposures to artificial sea water proved to be the minimum time required to achieve quantitative results (eliminate adaptation effects) with antennular receptrodes.

A significant problem associated with antennular receptrodes is lifetime. Receptrodes currently remain viable for approximately 48hrs.   This lifetime must be extended if more thorough studies are to be performed.   Decapod motor axons have been reported to remain physiologically active in vivo for 250 days after being severed from their cell bodies.(12)   A possible approach to extending the lifetime of antennular receptrodes is the duplication of these *in vivo* conditions *in vitro*.

Another difficulty arises because of the presence of tactile receptors in the antennules. These receptors, which respond to pressure changes, can mask the action potentials generated by the chemoreceptors.   Although the computerized data sorting technique provides a means of discrimination between the tactile and chemoresponses, care must be taken to minimize the tactile responses to limit the amount of data sorted by the computer, thereby increasing the calculation time. The design of the flow cell minimizes pressure changes by providing a smoothly flowing carrier stream.

Conclusions:

In contrast to most biosensors, in which only the molecular recognition element is of biological origin, antennular receptrodes incorporate biological components as both molecular recognition elements and transducers.   The complex multi-unit data generated by these biosensors can be analyzed by employing amplitude sorting algorithms similar to those used in clinical encephalography. These programs are relatively basic and can be run on most personal computers.

Generally, antennular receptrodes, or "neuronal biosensors", possess many desirable characteristics: short response times, high degrees of specificity and sensitivity, broad response ranges, and the ability to respond to various types of analytes. The chemoreceptors employed are not species limited, that is, chemoreceptors from several species have been successfully utilized to construct antennular receptrodes.   All of the sensors to date are of conveniently small dimensions, with sensing tips measuring approximately 1mm^2.   Although the system is not yet a practical sensor; as a model, the system demonstrates the vast potential of neuronal-based biosensors.

Literature Cited

1. Buch, R.M.; Rechnitz, G.A. *Biosensors*, 1989, 4, 215.
2. Buch, R.M.; Rechnitz, G.A. *Anal. Lett.*, 1989, 22, 2685.
3. Gleeson, R.A. *Biol. Bull.*, 1982, 163, 162.
4. Ghiradella, H.; Cronshaw, J.; Case, J. *Protoplasma*, 1968, 66, 1.
5. Ghiradella, H.; Case, J.; Cronshaw, J. *Am. Zoologist*, 1968, 8, 603.
6. Stanford, A.L. *Foundations of Biophysics*, Academic Press, New York, 1975; Chapter 5.
7. Cavanaugh, G. *Formulae and Methods*, Marine Biological Laboratory, Woods Hole, MA, 1964; 50.
8. Mulloney, B.; Selverston, A. *J. Comp. Physiol.*, 1984, 155, 593.
9. Buch, R.M.; Rechnitz, G.A. *Anal. Chem.*, 1989, 61, 533A.
10. Lockwood, A.P.M. *Aspects of the Physiology of Crustacea*, Freeman, San Francisco, 1967; 250.
11. Thompson, M.; Krull, U.J. *Anal. Chem. Symp. Ser.*, 1986, 25, 247.
12. Kennedy, D.; Bittner, G. *Cell Tissue Res.* 1974, 148, 97.

RECEIVED May 27, 1992

Chapter 6

Immunoelectrodes for the Detection of Bacteria

Judith Rishpon, Yigal Gezundhajt, Lior Soussan, Ilana Rosen-Margalit, and Eran Hadas

Department of Biotechnology and Molecular Microbiology, Faculty of Life Science, Tel-Aviv University, Ramat-Aviv 69978, Israel

This work describes a rapid and sensitive method for the determination of bacteria. The method is based on an enzyme-tagged immuno-electrochemical assay. Antibodies are immobilized on disposable carbon felt disc electrodes and are used to capture antigens in test solutions. After a short incubation with a second antibody, which is labeled with the enzyme alkaline phosphatase, the activity of the enzyme electrode thus formed is measured. This enzyme reacts with the substrate p-aminophenyl phosphate and the product of this enzymatic reaction, p-aminophenol, is detected amperometrically. The use of rotating electrodes significantly shortens the incubation times and, together with the computerized electrochemical system, results in extremely high sensitivity. The results obtained with *Staphylococcus aereus* and *Escherichia coli* cells show that the system can detect as low as 10 cell/ml in less than 10 minutes.

There is a great need for fast and reliable devices capable of detection and identification of various types of bacteria. Traditional microbiological methods are based on cultural techniques which are tedious and slow. A considerable effort has been invested in development of detection methods based on immunological interactions. These methods, which generally involve labeling of antibodies with radioactive, fluorescent or enzymatic label, have been shown to be quite effective, giving specific and quantitative detection of target antigens. During the last decade several attempts were made to combine the specificity of antibody- antigen interaction with the high sensitivity, wide dynamic range, and simplicity of electroanalytical methods. These attempts led to the development of electrochemical immunosensors, several of which detect the corresponding antigen at extremely high sensitivity (1-5).

0097–6156/92/0511–0059$06.00/0

In electrochemical enzyme immunoassays, an antigen or an antibody is ordinarily tagged with an enzyme and the enzymatic reaction is monitored by a potentiometric or amperometric electrode (6-13). The amplification obtained by enzyme catalysis is particularly advantageous for the detection of very low concentrations (14-15). A semihomogenous amperometric immunsensor for protein A- bearing Staphylococcus aureus in foods has been recently reported by Mirhabibollahi et al. (16). These authors report a quantitative assay in the range 10^4 to 10^7 cfu/ml. Recently, some reports have described the use of an electrode surface as both the immunological solid phase and as the electrochemical detector (17-19). In a previous publication we have described preliminary results obtained with an immunsensor based upon the enzymatic reaction of alkaline phosphatase with p-aminophenyl phosphate (20). In that system, glassy carbon served both as the solid phase in the immunorecognition reaction and as an amperometric electrode for oxidation of p-aminophenol formed by the enzymatic reaction. Glassy carbon electrodes were also used in a heterogeneous immunoassay using glucose oxidase as an enzyme label (17-18). By using a rotating disc electrode, Huet et al. (21) have recently shown that mass transfer to the solid phase, consisiting of antibodies immobilized on the glassy carbon, is a key step in heterogeneous immunassay.

In a preliminary report (Hadas, E., Sousan, L., Rosen-Margalit, I., Farkash, A. and Rishpon J., J. Immun. Assay, paper submitted) we have shown that by varying the nature of the carbon surface and increasing the number of the antibody binding sites we could considerably extend the detection limit of the electrodes. This was made possible by the use of electrodes made of carbon felt and by the covalent binding of the antibodies to the carbon surface via a spacer. The importance of a spacer arm in ligand coupling has been documented by several reports (22). An adequate spacer promotes the effective utilization of active sites with a minimum degree of blockage and allows for flexibility and mobility of the antibody molecule as it protrudes into the solvent. In this work the specific sequence employed was to bind biotin via an aliphatic spacer (hexamethylene-diamine) to the electrode surface, and use the biotin end to form a biotin/avidin/biotin bridge to the antibody. The use of a biotin /avidin/biotin bridge is a well known strategy in immunochemistry (23-24), and lately the same bridge has been applied to bind an enzyme to an electrode surface (25). The present paper describes the application of that system to the detection of extrememly low concentrations of bacteria.

Materials and Methods

Materials. 1-Cyclohexyl-3-(morpholino-ethyl)carbodiimide metho-p-toluenesulphonate (CCD), avidin, N-hydroxy-succinimide biotin (NHS-biotin) and Protein A from S. aureus (Cowan strain) were obtained from Sigma (USA). 1, 6 Diaminohexane, (HMD) was obtained from Fluka AG (Germany). P-aminophenyl phosphate (APP) was synthesized as described previously (26). Biotin rabbit anti-mouse IgG, alkaline phosphatase-conjugated rabbit antimouse IgG, and alkaline phosphatase - conjugated goat anti-rabbit IgG were obtained from Jackson Immunoresearch Laboratories (USA). Formalin-fixed Staphylococcus aureus (Staph. A) was obtained from Sigma.

Escherichia coli HB101 (a non pathogenic strain of *E. coli*) was obtained from M. Greenberg of the Department of Biotechnology and Microbiology, Tel Aviv University. Carbon felt sheets (RVG 1000) were obtained from Le Carbon Lorraine (France). Carbon felt discs (5 mm diameter) were cut from the carbon felt sheets.

Preparation of rabbit α-*E. coli* antibodies. 1 ml of 1.5 x10^7 HB101 cells together with 1 ml complete Freund's adjuvant were injected into rabbits. Injections were repeated four times every 3 weeks. However, after the first injection, 1 ml of incomplete Freund's adjuvant was substituted for the complete adjuvant. Blood from the rabbit ear was then collected in a test tube, kept 1 hour at room temperature and stored at 4°C overnight. It was then centrifuged and the antibody level of the supernatant was determined using standard ELISA techniques. The antibodies were precipitated by addition of ammonium sulphate at 4°C, followed by centrifugation at 10000 rpm. The precipitate was then dissolved in PBS and dialyzed in PBS for 48 hours at 4°C.

Preparation of mouse α *E. coli* antibodies. The method used was similar to that used with the rabbit α E. coli, but the volume injected was 0.5 ml and the concentration of the bacteria cell was 1.5x10^6 per injection. The injections were repeated once a week and the blood was drawn from the mouse eye.

Biotinilization of the Antibodies. 0.1ml of NHS-biotin in DMSO was added to a solution of 1mg/ml antibody (affinity purified rabbit antimouse IgG or anti HB101 prepared in rabbits) in 0.1M, pH 8, carbonate buffer. The mixture was stirred at room temperature for 4 hours, dialyzed against PBS at 4°C, centrifuged, and the supernatant collected.

Immobilization of Antibodies. The carbon felt discs were immersed in a solution of 1M HMD and 500 mg/ml CCD in water, the pH was adjusted to 5 (with HCl),   and the discs were incubated for 16 hours at ambient temperature with shaking followed by a thorough washing with PBS. The discs were then immersed in a solution containing 0.1M, pH 8, phosphate buffer 1 mg/ml   N-hydroxy-succinimide-biotin   (previously   dissolved   in dimethylformamide) and incubated for 16 hours at ambient temperature with shaking. The discs were then washed well with PBS, immersed in a solution containing avidin 50 μg/ml avidin in PBS and incubated for 10 minutes with shaking. Finally the discs were immersed in a solution contained PBS and the biotilinated antibody. In the case of *Staph. A.*, biotinilated rabbit antimouse Ig antibodies were used, while in the case of *E. Coli*, the biotinilated antibodies prepared were used. The immersed discs were incubated for 10 minutes with shaking and washed extensively with PBS.

Immunoelectrochemical Assay.   Prior to use the carbon felt discs were mounted on a housing made of a teflon cylinder containing a concentric platinum wire and a teflon cap with a stainless steel mesh (Figure 1). The carbon felt disc mounted in this assembly served as both heterogeneous phase for antigen (analyte) capture and as the working electrode in the amperometric measurement. The assay was performed with 5-10 electrodes simultaneously.

The immunoelectrochemical assay consists of three steps:

1. **Bacteria Capture:** Each of the antibody coated electrodes was immersed in 1 ml of the antigen *(Staph. A*, protein A or *E. coli* at different concentrations) and rotated at about 1300 rpm for 5 minutes (unless otherwise specified), followed by gentle washing in PBS.

2. **Conjugate binding:** For the analysis of *Staph. A* and protein A, the electrodes were then introduced into a solution containing AP conjugated rabbit antibody (Sigma AP rabbit antimouse IgG, diluted 1:250 2%), Tween 20 and 1% BSA in PBS, and rotated for 5 minutes at 1000 rpm. For the *E. coli* analysis, the electrodes were introduced into a solution containing mouse or rabbit IgG anti HB101, together with 2% Tween 20 and 1% BSA in PBS. The electrodes were then rotated at 1300 rpm (unless otherwise specified) for 5 minutes, washed gently in water and introduced into a solution containing AP rabbit anti-mouse IgG or AP conjugated goat anti-rabbit and rotated for 5 minutes at 1300 rpm.

3. **Electrochemical determination of enzyme activity:** After being gently washed in water the electrodes were transferred into the electrochemical cell containing 5 ml of substrate solution (3.7 mg/ml) of PAPP in 50 mM carbonate buffer (pH 9.6) and rotated at about 500 rpm. The PAR 273 potentiostat was used for the electrochemical measurements. The antibody electrode was employed as the working electrode and a standard calomel electrode and a platinum mesh were used as the reference and counter electrodes, respectively, in an amperometric measurement. The working electrode potential was held at 0.22 V vs. SCE. A computerized electrochemical system described earlier *(27)* was employed for current reading and signal averaging. The system was capable of simultaneous readings of several working electrodes in the same substrate solution, using a single reference electrode, a single counter electrode and a multiplexer *(28)*. The electrochemical cell was thermostated at 37°C unless otherwise specified. Measurements were automatically repeated until readings were stabilized, which usually occurred within 1 minute.

A complete scheme of the immunelectrodes is presented in Figure 2.

Results

Detection of protein A bearing bacteria: Figure 3 presents results obtained with the immunoelectrodes for the determination of the bacteria *Staph. A*. As a comparison, the same immunoenzymatic system was employed in a regular ELISA test. The results of the latter assay are shown in Figure 4. It is evident

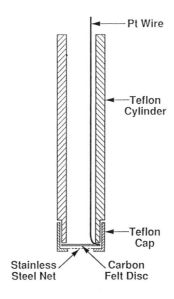

Figure 1. Schematic layout of the carbon felt electrode configuration.

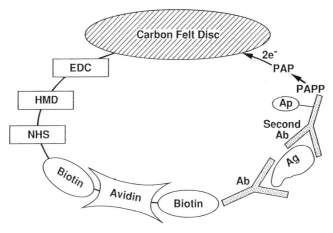

Figure 2. Scheme showing the immunoelectrode structure.

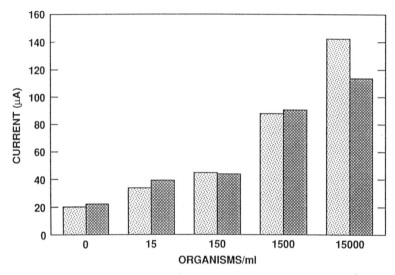

Figure 3.   Detection of *Staph A*. Amperometric response of the immunelectrodes after incubation with different concentration of the bacteria. The dark and light bars represent two sets of measurements.

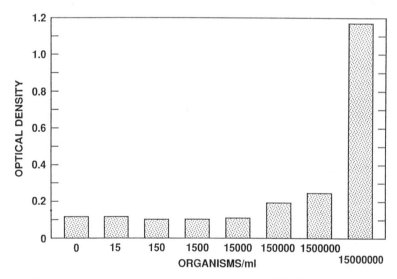

Figure 4. Detection of *Staph. A* by standard ELISA technique.

that the ELISA assay can detect only concentrations higher than 10^4 while the electrochemical immunoassay can detect as few as 10-100 bacteria.

The electrochemical step in the immunelectrodes is based on the detection of PAP which is oxidized at relatively low anodic potential (0.220V vs. SCE). The PAP is produced enzymatically from PAPP by the enzyme alkaline phosphatase (AP). The enzyme is attached to the carbon electrode via the rabbit IgG which is captured by *Staph. A* cells. The antigenic determinant, specific for the *Staph. A* organisms, is protein A, a protein present in the cell wall. We have studied the electrochemical characteristics of these electrodes and compared them with electrodes in which the bacterial antigen was replaced by the pure protein A. The cyclic voltammograms obtained are presented in Figures 5 and 6. Figure 5 shows the curves obtained for the capture of the *Staph. A*, whereas Figure 6 shows curves obtained with the protein A. The appearance of the PAP oxidation peak is clearly seen in both cases. However, in the presence of the bacteria the oxidation peak is slightly shifted to more anodic potentials, possibly due to mass transport limitations caused by electrode fouling by the bacteria.

Effect of temperature: Temperature might influence the immunoelectrochemical measurement in three ways: 1. by affecting the binding kinetics of the analytes to the immobilized antibodies and the binding kinetics of the conjugates; 2. by affecting the kinetics of the enzyme action; 3. by affecting the electrochemical kinetics. In a previous publication (Hadas, E., Sousan, L., Rosen-Margalit, I., Farkash, A. and Rishpon J., J. Immun. Assay, paper submitted) we showed that increasing the temperature in the electrochemical cell from 24°C to 37°C resulted in higher currents which improved the sensitivity by a factor of 2. In this work we have explored further the effect of the temperature. Figure 7 shows the effect of the temperature of

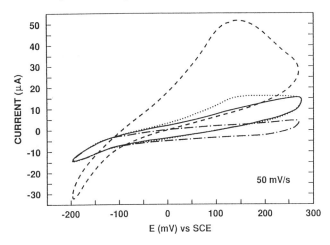

Figure 5. Cyclic voltammograms of *Staph. A* immunoelectrode.
·-· : buffer only blank; —: buffer and PAPP; ···: buffer, PAPP and 1.5×10^1 bacteria cells; --- : buffer, PAPP and 1.5×10^2 bacteria cells.

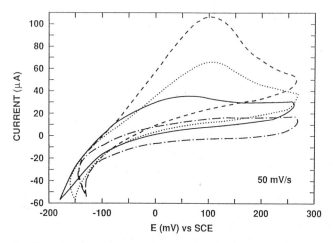

Figure 6. Cyclic voltammograms of *protein A A* immunoelectrode.
·-· : buffer only blank; ——: buffer and PAPP; ·····: buffer, PAPP and 1x10⁻¹³
gram/ml protein A; --- : buffer, PAPP and 1x10⁻¹² gram/ml protein A.

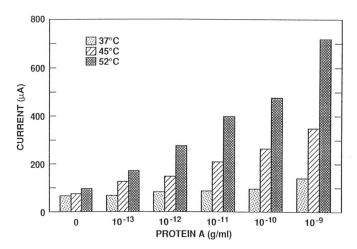

Figure 7. Effect of temperature on the electroenzymatic detection of protein
A.

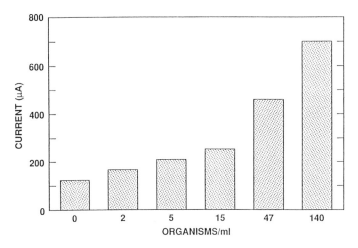

Figure 8. Detection of *Staph A* at 52°C.

the electrochemical cell on the signal obtained in the determination of protein A. Much higher sensitivity is obtained at 52°C. It should be noted, however, that the higher temperatures cause decomposition of the substrate PAPP and hence the electrochemical measurement has to be completed with 2-3 minutes. Further increase of the temperature resulted in rapid decomposition of the substrate and thus precluded a reliable measurement.

Results obtained with *Staph. A* at 52°C are shown in Figure 8. The figure shows that an order of magnitude increase in the sensitivity is obtained at the highest temperature used, 52°C, over that seen at 37°C.

Effect of electrode rotation: The antigen -antibody reaction in solution is usually not diffusion-limited. However, reactions at the solid/liquid interface are limited in practice by mass transport *(29)*. A rotating electrode causes a flow of fluid perpendicular and tangential to the electrode surface, resulting in the formation of a diffusion layer of constant thickness *(21)*. It has been previously shown that the flow of antigen molecules to the solid electrode surface is controlled by the parameter D/δ i.e. by the diffusion coefficient (D) of the antigen and by the thickness of the diffusion layer (δ). We have recently shown (Hadas, E., Sousan, L., Rosen-Margalit, I., Farkash, A. and Rishpon J., J. Immun. Assay, paper submitted) that rotation of an electrode made of carbon felt, which is non-rigid and has a large surface area, in a finite volume of the antigen solution, effects a rapid contact between the solution and the surface. Moreover, when the immobilized antibodies are present at sufficiently high concentration, they efficiently capture a high proportion of the antigen and consequently the incubation time required is significantly reduced. In the system described in this work, there are five steps in which the rotating electrode is involved: 1. avidin binding; 2. binding of the biotinilated antibody; 3. antigen binding; 4. binding of the AP conjugate antibody; 5. electrochemical determination of the enzyme activity. The last step is independent of the rotation speed as has been shown before *(20)*. Figure 9 shows the results obtained using two different rotation rates for the different heterogeneous immunoassay steps. The data shown demonstrates that the step which is most affected by the rotation is the binding of the first antibody. It further shows that high concentration of the immobilized antibodies leads to higher sensitivity and is a key factor in the fast antigenic reaction.

Detection of HB101. The antibodies prepared in rabbits and mice against the HB101 *E. coli* bacteria were used in the determination of the bacteria employing the "double sandwich technique". After bacteria capture by the rabbit anti HB101 antibodies, two schemes were used. In the first scheme, the same anti-HB101 rabbit IgG was used to attach to the bacteria. This was followed by an AP conjugated goat anti-rabbit IgG capture. In the second scheme, the mouse anti-HB101 IgG was employed as the second antibody, followed by capture of the AP conjugated rabbit anti-mouse. The electroenzymatic reaction was performed at 52°C. Figure 10 presents the results obtained using these two schemes. The results obtained by employing the first scheme are shown in the lower curve where the upper curve shows the reslts obtained by employing the second scheme. Both schemes result in detection limit of 10-100 bacteria/ml. Although the overall currents were higher in the first scheme, which employed the same antibodies on both sides of the sandwich, the total sensitivity of the measurement was lower compared to the second scheme, which employed different antibodies on both sides of the

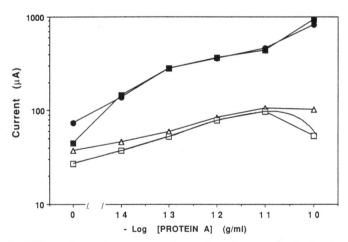

Figure 9. Effect of rotation rate on the various immunological steps. Empty squares: all steps at 1300 rpm; empty triangles: Biotinilated antibody binding at1300 rpm, other steps at 2600; full squares: avidin binding and AP-IgG binding at 1300 rpm, other steps at 2600 rpm, full circles: protein A binding at 1300 rpm, other steps at 2600 rpm.

sandwich. The latter scheme results in higher specificity, less non-specific adsorption and higher accuracy.

Detection Of Bacteria in Milk. The assay described was further tested in milk samples. Bacteria at different concentrations were added to the milk (3% FAT UHP) and the procedure described was employed for their detection. The results obtained for the *Staph. A* and for the HB101 *E. coli* are presented in Figure 11. The data clearly show that the immunoelectrodes developed could be applicable in fast identification and quantitation of bacterial contamination in food.

Discussion

The use of immunological techniques for the identification and quantitation of bacteria has received considerable attention lately. There is great interest in increasing the sensitivity of the immunoassay and in the development of fast, simple and reliable methods. However, rapidity and ease of operation are very often traded-off against high sensitivity. Indeed, in most cases described, some pre-enrichment step, usually of 18 hours duration, is essential to increase the number of target organisms to a detectable level (30-31). The immunelectrodes described in this paper are capable of direct detection of very low numbers of organisms, employing a simple, automatable and rapid procedure. This was made possible by the use of the carbon felt as the solid phase coupled with the high concentration of the antibodies bound covalently to the surface via a spacer. The large capacity of the solid phase results in the high sensitivity and wide dynamic range obtained. The use of rotating electrodes during the immunological reaction enables the solid phase to be rapidly exposed to all constituents of the solution. Consequently, incubation times are considerably reduced, down to several minutes. Moreover, the sensitivity of the electroenzymatic reaction is significantly increased by the elevated temperatures employed. Hence, the immunelectrodes described in this paper present a highly sensitive immunoassay.

Since the steps prior to the antigen capture can be carried out in advance, a large number of carbon felt discs coated with the first antibody (see figure 2) can be prepared and stored. Hence, the actual measurement consists only of capture of the antigen and the second antibody, followed by the electrochemical measurement. All of these steps can be performed with a large number of electrodes simultaneously. The latter aspect is significant particularly to clinical and other laboratories dealing with large number of antigen samples.

The rotation of the electrodes poses some mechanical requirements. Another strategy that might be adopted is the use of flow-through electrodes, in which the carbon felt is used as a column through which the analyte flows. Such experiments are now being conducted in our laboratory and the results seem promising.

The immunelectrodes described here are disposable. Although much effort has been invested in the development of reusable immunosensor, we believe that the low cost of the carbon felt and the tiny amounts of the antibodies used are in favor of disposable electrodes. This approach is

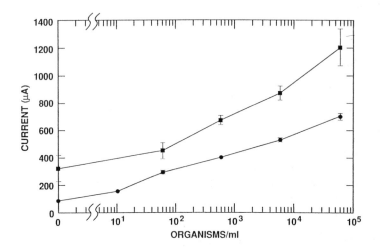

Figure 10.   Detection of *E. Coli.*

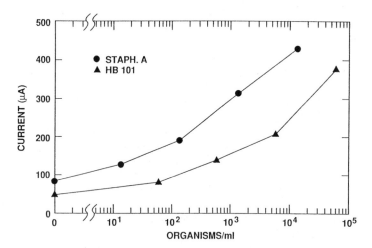

Figure 11.   Detection of bacteria in milk.

especially advantageous in the detection of pathogenic bacteria and also in measurements done in body fluids.

The procedure described for the determination of the *E. coli* bacteria contains also the antibodies preparation step and, hence, demonstrates the possible adaptation of the immunoelectrode system to various other types of bacteria and to other antigens as well.

Literature Cited

1. Janata, J. *J. Am. Chem. Soc.,* 1975, *97*, 2914,
2. Ngo, T. T. *Electrochemical Sensors in Immunological Analysis.* Plenum Press; New York, 1987.
3. Heineman, W.R. and Halsall, H.B. *Anal. Chem.* 1985, *57*, 1321(A).
4. Libby, J.M., Wada, H.G. *J. Clin. Microbiol.* 1989, *27*, 1456.
5. Czabon, J.D. *Anal. Chem.* 1985, *57*, 345A.
6. Heineman, W.R., Andersen, C.W. and Halsall, H.B. *Science.* 1979, *204*, 865.
7. Gebauer, C.R. and Rechnitz, G.A. *Anal. Biochem.* 1982, *124*, 338.
8. Alexander, P.W. and Maltra, C. *Anal. Chem.* 1982, *54*, 68.
9. Ngo, T.T., Bovard, S.H. and Lenhoff, H.M. *Appl. Biochem.Biotechnol.* 1985, *11*, 63.
10. Boitieux, J.L., Romette, J.L., Aubry, N. and Thomas, D.A. *Clin. Chim. Acta.* 1984, *136*, 19.
11. Suzuki, S. and Karube, I. in *"Applied Biochemistry and Bioengineering"*, L.B. Wingard, Jr., E. Katchalski-Katzir and L. Goldstein (eds.), Academic Press, London, 1981, vol 3, pp 145-174.
12. Whemeyer, K.R., Halsall, H.B. and Heineman, W.R. *Clin. Chem.* 1985. *31*, 1546.
13. Whemeyer, K.R., Halsall, H.B., Heineman, W.R., Voll, C.P. and Chen, I.W. *Anal. Chem.* 1986, *58*, 135.
14. Jenkins, S.H., Heineman, W.R. and Halsall, H.B. *Anal.Biochem.* 1988, *168*, 292.
15. Stanley, C.J., Cox, R.B., Cardosi, N.F. and Turner, A.P.F. *J. Immunol. Methods.* 1988, *112*, 153.
16. Mirhabibollahi, B., Brooks, J.L. and Kroll R.G. *App. Microbiol.Biotechnol.,* 1990, *34*, 242.
17. Robinson, G.A., Cole, V.M., Rattle, S.J. and Forrest, G.C. *Biosensors* 1986, *2*, 45.
18. Gyss, C. and Bourdillon, C. *Anal. Chem.,* 1987, *59*, 2350.
19. Rishpon, J. and Rosen, I. *Biosensors,* 1989, *4*, 61.
20. Rosen, I. and Rishpon, J. *J. Electroanal. Chem.* 1989, 258.
21. Huet, D., Gyss, C. and Bourdillon, C. *J. Imm. Methods* 1990, *135*, 33.
22. Hou, K.C., Zaniewski, R. and Roy S. *Biotechnol. and Appl. Biochem.,* 1991, *13*, 257.
23. Wilchek, M. and Bayer, E.A. *Anal. Biochem.* 1988, *171*, 1.
24. Griffin, J. and Odell, W.D. *J. Imm. Methods* 1987, *103*, 275.

24.   Griffin, J. and Odell, W.D. *J. Imm. Methods* 1987, *103*, 275.
25.   Pentano, P., Morton, T. H. and Kuhr, W.G. *J. Am. Chem. Soc.* 1991, *113*, 1832.
26.   Boyland, E. and Manson D. *J. Chem. Soc.* 1957, 4689.
27.   Rishpon, J. *Biotechnol. & Bioengin.*, 1987, *XXIX*, 204.
28.   Rishpon, J. In *"Biotechnology - Bridging Research and Application"*, Kluer Academic Press 1991.
29.   Stenberg, M. and Nygren, H. *J. Imm. Methods*, 1988, *113*, 2.
30.   Mirhabibollahi, B., Brooks, J.L. and Kroll, R.G. *J. Appl. Bacteriol* 1990, *68*, 577.
31.   Mirhabibollahi, B., Brooks, J.L. and Kroll, R.G. *Lett. Appl. Microbiol.* 1990, *11*, 119.

RECEIVED May 6, 1992

Chapter 7

Drug Detection Using the Flow Immunosensor

Frances S. Ligler, Anne W. Kusterbeck, Robert A. Ogert,
and Gregory A. Wemhoff

Center of Bio/Molecular Science and Engineering, Naval Research
Laboratory, Ser 6090.1-/255, 4555 Overlook Avenue,
Washington, DC 20375–5000

A detection system for small molecules has been developed which relies on the ability of antigen to displace fluorescent labelled antigen from antibody under flow conditions. The immunosensor responds in under a minute and is sensitive to pmoles of antigens. Antibodies to cocaine are incorporated into this system to exemplify its operational features.

Introduction to the Flow Immunosensor

The demand for a fast, specific, sensitive method for screening clinical and environmental samples for small molecules is high. Detection of pollutants, workplace hazards, explosives, therapeutic drugs and drugs of abuse are areas where such a system will find use. We have developed a highly sensitive, antibody-based sensor that detects ng (pM) quantities of small molecules in less than a minute.

Most antibody-based detection methods rely on either direct binding of antigen (i.e. sandwich immunoassays) or a competitive binding of antigen versus labelled antigen. Only the latter configuration has been widely used for the detection of small molecules. The flow immunosensor relies on a distinctly different antibody-antigen interaction. All of the antibody binding sites are saturated with a fluorescently labelled analog of the antigen prior to the immunoassays. When antigen is introduced into the system under nonequilibrium conditions, the fluorescent antigen is displaced from the antibody within seconds and flows into a detector.

A cartoon of the flow sensor is shown in Figure 1. We have developed the flow immunosensor for the detection of cocaine by immobilizing antibodies

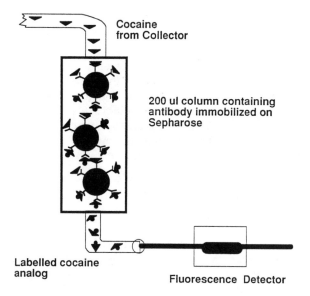

Figure 1. Schematic of the Flow Immunosensor.

to benzoyl ecgonine, the major metabolite of cocaine, on the surface of Sepharose beads *(1)*. The antibody-coated beads are exposed to fluorescein cadaverine benzoyl ecgonine at an antigen-antibody ratio of 100:1 to saturate the antigen-binding sites of the antibodies. The beads are then placed in a 200 ul column, approximately 5 nm long by 3 mm in diameter and a flow stream is established through the column. As cocaine is introduced into the flow stream, it enters the column and displaces the labelled antigen. The labelled antigen flows downstream through the flow cell of a simple fluorimeter and generates a signal. A laptop computer is used to control the valves and a small peristaltic pump and to save the fluorescence data.

Critical Parameters

Several factors critical to the performance of the flow immunosensor are listed in Table I.

TABLE I. Factors Critical to Performance

1.	Efficiency and speed of sampler
2.	Antibody affinity and stability
3.	Flow rate
4.	Monovalency of labelled antigen
5.	Nonspecific binding of labelled antigen
6.	Fluorescence of labelled antigen
7.	Interferents in the flow stream
8.	Sensitivity of fluorescence detector

The flow immunosensor can be used to detect antigen in either a pulsed or continuous stream. Thus it is readily adaptable for use with air samplers that extract vapors into water on a continuous basis, super sipper systems that rapidly inject samples from hundred of vials or microtiter wells, or individual samples injected by hand.

The antibody affinity is critical since labelled antigen bound too loosely may be washed off by the flow stream while antigen bound too tightly may never be displaced. In practice, the typical monoclonal antibodies, with moderate affinities (10^{-6} -10^{-9} 1Qmol^{-1}) in solution work well using flow rates of 0.1 - 1.0 ml/min. Within this range, the higher the affinity the higher the flow rate that produces optimum signals. (We have not tested antibodies with higher solution affinities.) The most important consideration with regards to affinity may actually prove to be the difference in affinity of the antibody for the antigen versus the labelled antigen. The greater the difference, the more efficient should be the displacement reaction.

The fluorescent antigen itself has several other critical characteristics. First, it must have only one epitope for antibody binding. If the fluorescent antigen were bound by more than one epitope and only one epitope was displaced from the immobilized antibody, the flourescent antigen would not be released into the flow stream. Thus a higher concentration of antigen would be required for displacement and the sensitivity of the sensor would be reduced. Second, the labelled antigen must not bind nonspecifically to the column materials. Particular care must be taken when using fluorophores and antigens which are hydrophobic. One solution to this problem is to couple one antigen molecule and fluorophores to the hydrophilic carrier to keep the complex in solution (2). This solution also addresses the third important feature of the labelled antigen. The signal generated as a labelled antigen is displaced is directly proportional to the number of fluorophores coupled to the antigen. The fluorescein cadaverine benzoyl ecgonine used in the studies described here has a fluorophore-to-antigen ratio of 1:1. The carrier system tested in the flow immunosensor for detection of dinitrophenol (3) has a fluorophore-to-antigen ratio of 3:1.

Interferents from the sample are capable of generating a false positive signal. Proteases, heat, or very low pH will denature the antibody and cause the release of labelled antigen. Fluorescent elements in the samples may mask fluorescence from released labelled antigen. For analysis of urine samples for cocaine metabolites, for instance, we found that many urines fluoresce at the same wavelengths as fluorescein. Switching to rhodamine labels and higher wavelengths minimizes this problem.

The sensitivity of the detector is critical for measuring the small numbers of displaced labelled antigen molecules. Fluorimeters designed for use with HPLC generally have the sensitivity required. The systems used in our laboratory are capable of detecting 10^{13} M fluorescein. A laser-based system could be designed for enhanced sensitivity, but so far the cost has outweighed the potential benefit.

Cocaine Detection

The performance of the flow immunosensor for the detection of cocaine was evaluated as flow rate, antibody density, and cocaine concentration were varied.

Figure 2 shows the signal generated when relatively large amounts of antigen (1:1 mole ratio antigen-to-antibody) were added repetitively to the column at varying flow rates. There is slightly more signal generated at the slower flow rates. However, all three flow rates produce readily detectable signals and the columns become depleted of displaceable labelled antigen and thus incapable of further signal generation after the same number of sample additions.

The sensitivity of the columns measuring small amounts of cocaine was also similar for flow rates of 0.5 to 1.0 ml/min. Figure 3 shows the fluorescence signal generated in 3 columns to which cocaine was added in

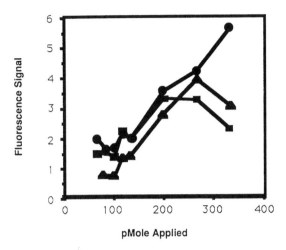

Figure 2. Effect of Flow Rate on Signal.
Flow rates of 0.5 ml/min (●), 0.75 ml/min (▲) and 1.0 ml/min (■) were established through a column containing a low density of immobilized anti-benzoyl ecgonine antibody saturated with fluorescein cadaverine benzoyl ecgonine. Cocaine was added to a column at a molar ratio of 1:1 antigen-to-antibody in repeated samples. The fluorescence from displaced labelled antigen was measured downstream using a Spectravision FD300 fluorimeter equipped with a 12 ul flow cell.

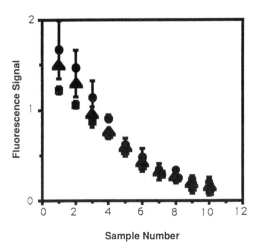

Figure 3. Affect of Flow Rate on Sensitivity
Increasing concentrations of cocaine were added to columns containing low density immobilized antibody saturated with labelled antigen. The column were run at flow rates of 0.5 ml/min (●), 0.75 ml/min (▲) and 1.0 ml/min (■).

amounts increasing from 50 pmoles to 330 pmoles. Interestingly, the columns run at faster flow rates were depleted of displaceable labelled antigen faster than the column run at 0.5 ml/min.

Since rebinding of displaced labelled antigen by immobilized antibodies is possible, an experiment was devised to examine the effect of antibody density on signal generation and displacement efficiency. Antibody was immobilized on Sepharose beads at 1.2 pMole/mg and 4.9 pMole/mg, saturated with 100-fold excess of labelled antigen, and placed in 200 ul columns. Repeated samples of cocaine at 10-fold molar excess over antibody were added to each column at the 0.5 nl/min. The slower flow rate was chosen in order to provide time for rebinding to occur. Figure 4 upper panel demonstrates a steady rate of depletion of labelled antigen in the column containing low density of antibody, but a slower rate of depletion in the column containing high density antibody.

In order to explore this issue further, higher antigen concentrations were added to the columns to prevent reassociation of displaced labelled antigen with antibody. Repetitive samples of 100-fold molar ratio antigen-to-antibody were added repeatedly to the column. Figure 4, lower panel, shows three things: First, as with low density antibody, the total signal is higher when high density antibody is used. This finding is not surprising since there is more antibody, more displaceable labelled antigen, and more free antigen in the assay. Second, the rate of displacement, as reflected in the slope of the lines, is the same for columns containing high and low density antibody. And third, the difference in the rate of depletion seen with low density antibody has disappeared. These findings suggest that rebinding of labelled antigen can occur. However, this phenomenon is more of a problem in columns containing high density of immobilized antibody and when the immunosensor is measuring small quantities of antigen.

In order to determine whether the fluorescence signal generated by the displacement reaction was a quantitative measure of the amount of antigen added, various concentrations of cocaine were added to previously unused columns. The columns contained a low density of antibody and were run at 1.0 ml/min. Figure 5 shows that the fluorescence intensity is directly proportional to the antigen concentration in the added sample. Furthermore, this relationship was true even in the low ng (pmole) ranges where the antigen concentration was too low to prevent rebinding of displaced labelled antigen, suggesting that maintaining low antibody density is sufficient to prevent rebinding of displaced labelled antigen. It should also be noted that slope of calibration line will change if column is used more than once. In order to maintain quantitative accuracy, columns must either be used only once! or the data must be normalized based on the amount of labelled antigen remaining on the column.

Conclusion

The flow immunosensor utilizes an antigen displacement reaction for the detection of small molecules. The sensor is fast, sensitive, and easy to operate.

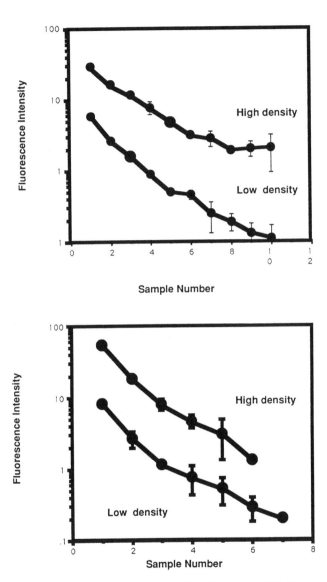

Figure 4. Effect of Antibody Density on Displacement Efficiency. Columns containing high or low density of anti-benzoyl ecgonine antibody saturated with fluorescein cadaverine benzoyl ecgonine were run at 0.5 ml/min. The fluorescence generated by the displacement of labelled antigen was measured following successive additions of cocaine at 10:1 molar ratio of antigen-to-antibody (upper panel) or 100:1 ratio (lower panel).

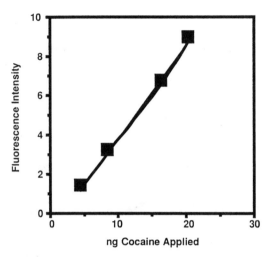

Figure 5. Sensitivity and Quantitation of Cocaine.
Aliquots of cocaine were added to unused columns containing low density anti-benzoyl ecgonine saturated with fluorescein cadaverine benzoyl ecgonine at a flow rate of 1 ml/min.

For the detection of cocaine, it is much lower cost and easier to operate than methods currently used in screening laboratories. Furthermore, the flow immunosensor can detect 5 ng of cocaine in less than a minute, which is sensitivity equivalent to radio-immunoassay.

Literature Cited

1.   Wemhoff, G.A.; Kusterbeck, A.W.; Ligler, F.S. *NRL Review* 1991, pp 103-105.
2.   Bredehorst, R.B.; Wemhoff, G.A.; Kusterbeck, A.W.; Charles, P.T.; Thompson, R.B.; Ligler, F.S.; Vogel, C.W. *Anal Biochem* 1991, *vol 193*, pp 101-108.
3.   Kusterbeck, A.W.; Wemhoff, G.A.; Ligler, F.S.; In *Biosensor Technology*; Buck R.P., et.al., Eds.; 25; Antibody-Based Biosensor for Continuous Monitoring; Marcel Dekker, NY, NY, 1990; pp 345-349.

RECEIVED July 15, 1992

Chapter 8

A Conductive Polymer-Based Immunosensor for the Analysis of Pesticide Residues

Robert G. Sandberg, Lisa J. Van Houten, Jerome L. Schwartz,
Robert P. Bigliano, Stephen M. Dallas, John C. Silvia,
Michael A. Cabelli, and V. Narayanswamy

Ohmicron Corporation, 375 Pheasant Run, Newtown, PA 18940

Testing for pesticide residues today requires highly trained personnel, sophisticated laboratory instruments, and time measured in days. The traditional approach utilizes GLC and HPLC technology and these require extensive sample preparation procedures that lead to high costs and long turnaround times. The development of an enzyme-linked immunosorbent assay (ELISA) that uses electroconductive polythiophene as the solid support (a biosensor) is described. Based on initial studies with electroconductive polyacetylene, the polythiophene biosensor is expected to reach sensitivities in the low parts per billion. Formats capable of detecting a range of pesticide residues in food and soil samples as well as in ground water are under consideration. A miniature field portable assay system is envisioned.

During the past 50 years, pesticides have been manufactured and used in increasing amounts throughout the world. Without question, they have contributed significant health and economic benefits to society. At the same time, however, wide-spread use of pesticides has created serious concerns regarding their effect on the environment. Particular attention has been focused on the impact of pesticide residues on drinking water and the food supply (1) and the pesticide industry now faces a worried public eager for scientific information and safety assurances. The general consensus is that the United States and foreign nations require increased testing programs. Besides helping professionals assess the situation regarding residues, increased testing most decidedly will help convince the public of the healthfulness of their drinking water and food supplies (2).

Until recently, the identification and quantitation of pesticide residues

0097–6156/92/0511–0081$06.00/0
© 1992 American Chemical Society

in ground water and other sources has been limited to traditional testing systems (such as gas chromatography, mass spectrometry, liquid chromatography, and thin-layer chromatography). These sophisticated systems require highly trained personnel. This, coupled with time consuming sample preparation procedures, largely restricts their use to a laboratory setting. Over the last several years, enzyme immunoassay technology (EIA), in wide use for many years in the clinical laboratory to measure human body fluid components, has been adapted for pesticide measurement. Immunoassays developed at Universities, those produced by agrichemical manufacturers for in-house quality control and commercial products, such as Ohmicron's magnetic particle based RaPID Assays® and ImmunoSystems tube and plate assays are rapidly growing as tools for initial screening. Positive samples are independently reconfirmed with traditional tests. This two-step approach minimizes the number of samples necessary to subject to more complex and costly analysis and significantly improves the efficiency of the testing laboratory. Experience with EIA diagnostic methodology in health care and drugs-of-abuse testing has demonstrated that the specificity, sensitivity, and general analytical quality of EIAs makes them ideally suited for this screening role. The current attention being given to immunoassay technology by the Food and Drug administration provides a clear signal of the growing acceptance of EIAs for environmental diagnostics (3).

EIA methods usually measure the intensity of color produced in a sequence of coupled reactions. The sequence typically involves a specific antibody bound to a solid support such as coated glass tubes, glass beads, microtiter plates, or superparamagnetic particles. The measurement and reporting of results involves wet chemistry procedures and a suitable spectrophotometer. In the assay, a purified enzyme-linked pesticide residue competes with the pesticide residue in the test sample for binding to a specific antibody that is attached to a magnetic particle or other support. The antibody-bound, enzyme-linked residue is separated magnetically from enzyme-linked residue that cannot bind to the antibody due to competition from the pesticide residue in the test sample. A colorimetric reaction catalyzed by the bound enzyme is measured photometrically. Results are usually available within 45 minutes.

Our work involving electroconductive polymer as the solid support provides an alternative to spectrophotometry. Utilizing analogous sequential coupled-reactions, a dopant (such as iodine) can be quantitatively produced by the bound enzyme. Under the influence of dopant, a suitably chosen electroconductive polymer can simultaneously serve as the solid phase to support immobilized antibodies and, through conductivity modulation by the dopant, function as the measuring device (4-6). The instrumentation required to measure electrical changes in such a system can be as simple as a voltmeter. Since the optical nature of the medium will not influence the measurement, this system is expected to be tolerant of colored or turbid samples. This will further simplify sample preparation. Particularly important, a miniaturized biosensor-based diagnostic will allow the fabrication of several different immunosensors side-by-side in one structure (such as a dipstick). This will enable the design of a single-use biosensor capable of simultaneous multiple analyte detection.

Polyacetylene-Based Biosensor

Three essential components are found in all biosensors: a bioactive surface that interacts with the substance to be measured; a transducer that detects the biochemical event occurring between the bioactive surface and the analytes; and support electronics that amplify and report the output signal from the transducer. One phase of our research focused on the development of a semiconductive polymer film sensor for glucose using the electroconductive polymer polyacetylene (7). We then created a suitable model sensor and initiated work to couple immunochemical reagents to the sensor.

Polyacetylene Sensor Synthesis. A Class 10,000 clean room was used for film synthesis and film-related material handling. Gas phase polymerization of acetylene within a processable open pore (spongy) polymer film was used to achieve a flexible polyacetylene matrix (Figure 1). This process utilized polyvinylidine fluoride filter media impregnated with a Ziegler-Natta catalyst. The catalyst saturated media was exposed to acetylene gas in a stainless steel reaction chamber equipped with special temperature and pressure control and monitoring capabilities (8). The resulting polyacetylene films were then predoped using a solution of iodine in toluene to adjust initial conductivity. This was followed by the application of a conductive layer of Electrodag™ (conductive screenable graphite mixture) through a patterned silkscreen. The finished films were punched into disks of approximately 11 mm diameter and mounted in specially designed cells (Figure 2). These single-use disposable cells functioned as a self-contained reaction chamber. Electrical connection to the cell was accomplished using spring-loaded contact pins to contact the Electrodag pattern through holes in the base assembly of the cell.

Polyacetylene Immunosensor Preparation. Atrazine antiserum was prepared by the method of Bushway (9). The antiserum was coated on the polyacetylene film sensor by direct adsorption in an 0.05 M phosphate buffer solution at pH 6.0 containing 5 M sodium chloride and 0.02 M potassium iodide. The coated films were then stabilized and dried using a proprietary method. The enzyme conjugate was obtained by covalently linking glucose oxidase to atrazine using a modified carbodiimide conjugation technique.

Bioassay Procedure. Resistance measurements were made using a pulsed voltage system multiplexed to monitor 16 cells, simultaneously. The cells were pulsed once every 38 milliseconds with an applied voltage adjusted to each cell's starting conductivity. With the cell in place, the voltage read across a 1 megohm load resistor was 0.35V. Pulse rate, cell selection, and data acquisition were controlled by an IBM PC-AT compatible computer using a Data Translation A/D D/A Interface Board DT2805. Cells were initially read for 1 minute to establish a baseline response. Then, after addition of a reagent to initiate the enzyme reaction, the potential across the load resistor was followed for an additional 2 minutes. Using ASYST Scientific Data Acquisition and Analysis Software, the data was collected and the slope of the voltage response versus time using a linear model was determined.

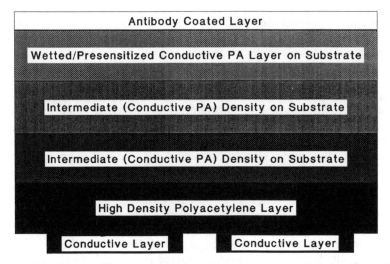

Figure 1. Sensor film cross-sectional structure revealed by Scanning Electron Microscopy.

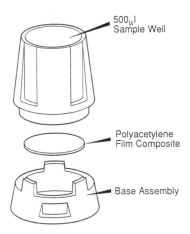

Figure 2. Single use 2-piece injection molded sensor cell.

The measurement of standard aqueous atrazine solutions was performed by adding 200 microliters of the solution to a coated biosensor. This was incubated for 5 minutes at room temperature. Then 25 microliters of the enzyme conjugate was added to each biosensor and the mixture was further incubated for 5 minutes at room temperature. The solution was decanted and the cell rinsed once with water. At this point, 200 microliters of 0.05 M phosphate buffer, pH 6.0 containing 0.15 M sodium chloride and 0.02 M potassium iodide was added to each sensor and they were placed on the pulse voltage system. The conductivity reading cycle was then initiated as previously described. At the end of the baseline period, 25 microliters of a solution containing 8 micrograms/mL of lactoperoxidase and 10 g/dL glucose in 0.05 M phosphate buffer, pH 6.0 with 0.15 M sodium chloride was added to each sensor. The reaction chemistry is summarized in Figure 3. At the end of the reading period, the results were calculated as described above and expressed as percent voltage change per minute. Responses from individual sensors were compared to the response of sensors used to analyze samples containing no atrazine. Typical dose response data is shown in Table I.

Table I. Dose Response Data for the Atrazine Biosensor Assay.

| | | Response | | |
Standard	N	Mean	S.D.	%B/Bo*
NSB	10	1.96	0.63	-
0	16	6.90	2.00	100.0
25 ppt	8	5.15	1.49	64.6
250 ppt	10	4.84	1.84	58.3
2.5 ppb	16	4.27	0.83	46.8
25 ppb	14	2.99	1.23	20.9
250 ppb	14	2.34	1.02	7.7

* Corrected for nonspecific binding

Analysis of the work done during this period has led to four important conclusions: Using an electroconductive organic polymer as both the antibody immobilization substrate and the transducer for a marker enzyme was practical; Optimization of the process of manufacture of Ohmicron's polyacetylene film substrate was complete; The nature of the process, the production of a polymer blend by gas phase polymerization of acetylene within an open-pore polymer film yielded, at best, a substrate having a precision of 15 to 30% C.V. Since

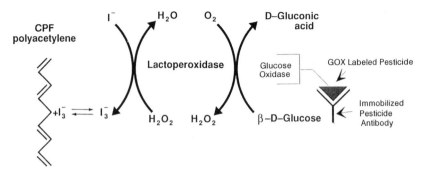

Figure 3. Labeled pesticide (from competition with unlabeled pesticide present in sample) is linked to the enzyme glucose oxidase. Detection is through a series of coupled reactions.

our goal for this stage of the program was a response precision of 10% or better, we initiated a search for other ways to create the electroconductive polymer support (sensor).

Polythiophene-Based Biosensor

The work with polyacetylene provided a rich foundation upon which to construct the next stage of the program and we have overcome the precision barriers encountered with the early intractable polymer-film blend process. Toluene soluble electroconductive formulations of polythiophene, spray coated onto the same open-pore polymer film substrates in which the polyacetylene was formed, yielded identical 15-30% performance. However, sensors produced from solutions of polythiophene, spincoated onto electrodes prepared photomechanically from gold coated Kapton® film has routinely produced sensors with response precision in the 6 to 10% C.V. range. Our progress over the first six months of this program is summarized in Figure 4.

The initial phase of this program was focused on the design and construction of the transducer (sensor). The best performing electrode pattern was selected and coated with polythiophene. Prototype immunosensors were then prepared by immobilizing antibody using the procedure described above for polyactylene and we successfully demonstrated a functioning biosensor. We are now identifying the most efficient route to immobilize and stabilize antibodies on the polymer surface and cooptimizing the coupled reagent system. A schematic of our design is shown in Figure 5 and the series of coupled reactions selected for product development are shown in Figure 6. In a parallel effort, building on our earlier system, we designed and assembled a multichannel instrument to measure sensor/biosensor response. This was interfaced to a personal computer to facilitate R&D and afford the flexibility to accommodate future changes in measurement algorithms, hardware, and computer platform. Preliminary software code has been written and implemented. Additional features and improvements will be added as our

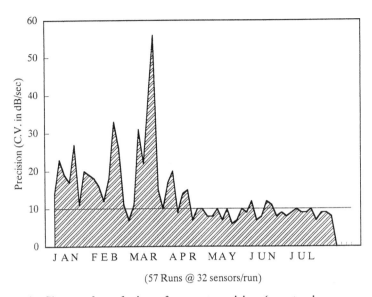

Figure 4. Six month evolution of sensor precision (process improvements).

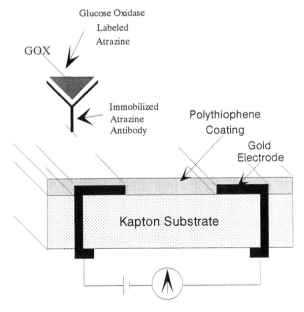

Figure 5. Immunosensor design schematic.

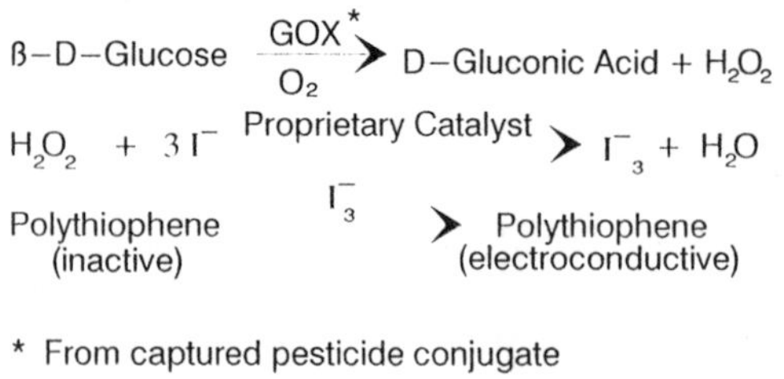

$$\beta-D-Glucose \xrightarrow[O_2]{GOX^*} D-Gluconic\ Acid + H_2O_2$$

$$H_2O_2 + 3\ I^- \xrightarrow{Proprietary\ Catalyst} I_3^- + H_2O$$

$$Polythiophene \xrightarrow{I_3^-} Polythiophene$$
$$(inactive) \qquad\qquad (electroconductive)$$

* From captured pesticide conjugate

Figure 6. Iodine dopant generated by three coupled reactions (bound GOX is the limiting reagent).

needs progress. We have used the same sensor cell and cell holder format previously designed for polyacetylene films to expedite the program. We are also using the same commercial antibodies developed for Ohmicron's RaPID® Assays ELISA kit. Further results of our program will be reported in 1992.

Literature Cited

1.   Petersen, B.;Chaisson, C. "Pesticides and Residues in Food," *Food Technology*, July, 1988, 59-64.
2.   Munch, D. J.; Maxey, R. A.; Engel, T. M. "Methods Development and Implementation for the National Pesticide Survey," *Environ. Sci. Technol*, **1990**, *24*, 1446-1451.
3.   Kaufman, B. M.; Clower, M., Jr. "Immunoassay of Pesticides," *J. Assoc. Off. Anal. Chem.*, **1991**, *74*, 239-247.
4.   Kanatzidis, M. G. "Special Report - Conductive Polymers," *C&EN*, December 3, 1990, 36-54.
5.   Malmros, M. K. "Analytical Device Having Semiconductive Polyacetylene Element Associated with Analyte-Binding Substance," U. S. Patent No. 4,334,880, June 15, 1982. (Assignee: **Ohmicron Corporation, DE**).
6.   Malmros, M. K. "Analytical Device Having Semiconductive Polyacetylene Element Associated with Analyte-Binding Substance," U. S. Patent No. 4,444,892, April 24, 1984. Continuation-in-part of Pat. No. 4,334,880. (Assignee: **Ohmicron Corporation, DE**).
7.   Malmros, M. K.; Gulbinski, J.; Gibbs, W. B., Jr. "A Semiconductive Polymer Film Sensor for Glucose," *Biosensors* **1987**, *3*, 71-87.
8.   Malmros, M. K.; Gulbinski, J. "Differential Homogeneous Immunosensor Device," U. S. Patent No. 4,916,075, April 10, 1990. (Assignee: **Ohmicron Corporation, DE**).
9.   Bushway, R. J.; Perkins, B.; Savage, S.A.; Lekouski, S. J.; Ferguson, B. S. *Bull. Eviron. Contam. Toxicol.* **1988**, *40*, 647-654.

RECEIVED May 6, 1992

Chapter 9

Application of Capacitive Affinity Biosensors
HIV Antibody and Glucose Detection

Herbert S. Bresler[1], Michael J. Lenkevich, James F. Murdock, Jr.,
Arnold L. Newman, and Richard O. Roblin

Biotronic Systems Corporation, 9620 Medical Center Drive, Suite 300,
Rockville, MD 20850

A miniature electronic biosensor system for the rapid detection of various analytes has been partially developed by Biotronic Systems Corporation (BSC). The system, called the Capacitive Affinity Sensor Instrumentation System (CASIS), represents an improvement over other rapid detection instruments. The present paper focuses on the CASIS as a system to detect antibodies to infectious disease organisms and small analytes such as glucose. As an antibody detector, CASIS provides more details of patient antibody status than conventional technologies in much less time while using significantly less patient blood. Tests for HIV-1 and Hepatitis B antibodies are currently under development. The antibody-detecting system was adapted to the detection of small antigens and sugars. Competition assays were developed to provide semi-quantitation of glucose and the antibiotic Gentamicin. A brief description of the sensor structure, its principles of operation, and a detailed description of the BSC Hydrocarbon Sensor System are provided as an introduction to the Biosensor. In addition to data on the detection and quantitation of hydrocarbons in water, data are shown for detection of HIV antibodies and glucose.

Current HIV antibody screening methods provide a general yes or no answer to the question, "does this individual have antibodies to HIV?" Since

---

[1]Current address: Cellco, Inc., Germantown, MD 20874

0097–6156/92/0511–0089$06.00/0

these tests are imperfect, positive results obtained in initial screening tests demand that the patient be re-tested by a confirmatory test called a Western Blot. A Western Blot provides detailed information about the presence or absence of antibodies to the several specific proteins of HIV, and is the definitive confirmatory test. Western Blots are expensive and time consuming. Typically a Western Blot costs $55 and takes about two days (Cambridge Biotech, Inc., Rockville, MD). The biosensor system described here provides some of the same protein-specific information currently obtained by Western Blot, but with speed and cost comparable to current initial screening tests.

Screening tests for antibodies to HIV are the primary means by which the spread of HIV infection is minimized. Screening for HIV contamination in blood or blood products for transfusion is an obvious opportunity to prevent the spread of HIV infection; about 20 million screening tests are performed for this purpose each year in the United States alone. These tests are currently performed after collection from the donor. A rapid test that requires only a drop of blood could be used to screen potential donors before donation, thereby avoiding the expenses and risks of collecting and disposing of unusable blood units.

Screening individuals at risk for AIDS provides another opportunity to prevent the spread of HIV infection by sexual or other person-to-person contact. However, current tests are not practical for many individuals because of high costs and delays in obtaining results. The Capacitive Affinity Sensor Instrumentation System (CASIS) provides the potential for development of a tabletop instrument that would provide specific antibody test results in less than 15 minutes, using only blood obtained by a finger-prick. Other advantages of the CASIS are:
  1) the instrument would conduct the entire test
  2) the instrument would report the results
  3) the simplicity of operation necessitates minimum
     training for an operator
  4) the test has an unambiguous end-point; that is, no
     operator interpretation is required.
Such an instrument would be suitable for pre-donation screening of blood donors and for use in a doctor's office or clinic.

Capacitive Affinity Sensor: Principles of Operation

The Capacitive Affinity Sensor (CAS) is a generic, solid-state, microelectronic device which is made specific for a particular analyte by coating the surface of the sensor with a material having affinity for that analyte. For example, specificity for antibodies is derived from covalently-linked antigens which selectively concentrate antibodies near the sensor surface. Figure 1 shows a schematic of the sensor structure. The sensor is a planar, interdigitated capacitor, the capacitance of which is modulated by the dielectric properties of substances within its electric field. Dielectric constant, an intrinsic property of all substances, is a measure of the ability of a substance to store energy in response to an electric field. The CAS takes advantage of the large difference in dielectric constant between aqueous buffers and other substances including proteins. Changes in the dielectric and/or conductivity of the material immediately adjacent to the sensor surface will induce a change in capacitance of the sensor. The principle of the Capacitive Affinity Sensor is the subject of patents in the United States, Canada and Europe. (Covered by one or more of the following patents: United States 4,728,822; 4,822,566; 5,057,430 - Europe 88904823.7 - Canada 1,256,944; 1,259,374 - and patents pending.)

Several strategies that apply specific coatings to take advantage of the capacitive principle have been developed. Different configurations of the CASIS may be utilized depending on the application. Although the instrumentation is somewhat different in each configuration, the basic components of each instrument are quite similar. The sensor is subjected to a low power, alternating current electric field, and the signal from the capacitor is received and processed by the instrument. The microprocessor component is designed to input the signal from the sensor and produce an output signal that is proportional to the capacitance change. The instrument has the ability to store the input data, and to transfer the stored data through its RS-232 serial interface to any IBM-compatible computer system.

Hydrocarbon Sensor Instrumentation System

A description of the BSC Hydrocarbon Sensor Instrumentation System is helpful to illustrate the performance of the capacitance principle. The specificity of this sensor is conferred by a polymeric surface coating that has the ability to capture and thus concentrate medium-chain aliphatic hydrocarbons such as those in gasoline and kerosene. Table 1 shows a partial list of the compounds tested for reactivity with the Hydrocarbon Sensor. The response of the sensor to n-hexane has been given the arbitrary value of unity, and all other compounds are shown relative to n-hexane. It is apparent from these data that aromatic hydrocarbons such as benzene are not detected by the sensor, nor are halogenated organics detected. The lack of reactivity minimizes interferences or false positive signals. A miscellaneous group of compounds including alcohols, ketones, acids, ethers, and amines were also tested with no interference. These data indicate that the Hydrocarbon Sensor, though not specific for a single chemical, is specific for a defined set of chemicals. Also included in Table 1 are data for unleaded gasoline and diesel

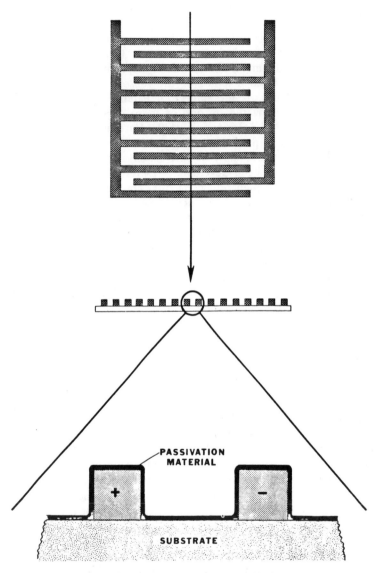

Figure 1. Schematic of capacitive affinity sensor.

fuel. Gasoline and diesel fuel are complex mixtures of many chemicals, but alkanes predominate. Petroleum products with a predominance other of light alkanes, such as gasoline, produce a stronger sensor response than diesel fuel or kerosene. Nevertheless, even complex fuels like diesel fuel with a large number of heavy hydrocarbon components still contain sufficient medium-chain alkanes to elicit a detectable response from the Hydrocarbon Sensor at low parts per million (ppm) levels.

Sensitivity and Dynamic Range. Figure 2 shows the linearity of the response of the Hydrocarbon Sensor when exposed to various levels of hexane. The dynamic range for hexane extends from a minimum detectable level of 200 parts per billion (ppb) to levels in excess of 20 ppm. Similar data have been generated for other alkanes. As most hydrocarbons have solubility limits in 25-50 ppm range, the Hydrocarbon Sensor can easily detect relevant concentrations of these materials in water, with coefficients of variation between runs of less than 5%.

Response Time and Reversibility. The response time of the Hydrocarbon Sensor is nearly instantaneous depending on the conditions under which it is operated. In Figure 3, the sensor response to an injection of 1.0 ppm hexane is indicated. Note that the sensor immediately responds to this concentration of hydrocarbon; the response attaining an equilibrium value within minutes. Also indicated is the reversibility of the Hydrocarbon Sensor. When the sensor is returned to hydrocarbon-free water, the sensor response returns to its original value within seconds. This rapid reversibility allows the Hydrocarbon Sensor to perform real-time, on-line hydrocarbon monitoring functions, such as effluent monitoring or process control.

The BSC Bubble Sensor

In the process of developing the Hydrocarbon Sensor, we observed tiny bubbles forming on the sensor surface, and concluded that bubble formation at the sensor surface was probably responsible for most of the observed change in capacitance in the Hydrocarbon Sensor system. When an air bubble was applied experimentally to a submerged sensor, this too resulted in a large change in capacitance. This seminal observation led to the development of a system based on the biochemical generation of bubbles. The HIV antibody detection system is one application of the Bubble Sensor.

The HIV Antibody Detection Sensor

The BSC Antibody Sensor system used specific antigens covalently applied to the sensor surface to confer antibody specificity to the sensor. Oxygen bubbles were generated by catalase reacting with H_2O_2. The schematic in figure 4 illustrates the basic experimental procedure. Amino-terminated silanes were covalently bound onto the SiO_2 passivation layer on the surface of a sensor chip. Conventional two-step glutaraldehyde fixation was used to covalently bind antigen to the silane. The sensor was

TABLE 1

APPROXIMATE RELATIVE SENSITIVITIES OF THE HYDROCARBON SENSOR
TO VARIOUS ORGANIC CHEMICALS AND FUELS

Substance	Relative Sensitivity
n-pentane	0.2
n-hexane	1.0
n-heptane	2.6
n-octane	5.4
n-nonane	5.2
2-methylbutane	0.1
2-methylpentane	0.8
3-methylpentane	0.7
2-methylhexane	2.4
3-methylhexane	2.2
2,2-dimethylbutane	0.4
2,4-dimethylpentane	2.0
cyclohexane	0.3
1-hexene	0.2
benzene	0.0
toluene	0.0
m-xylene	0.0
trichloroethylene	0.0
2-propanol	0.0
unleaded gasoline	1.6
diesel fuel #2	0.8

Absolute sensitivities are measured by adding 1 mg/l of the substances and recording the maximum response. Relative sensitivities are calculated by dividing this maximum response by the absolute sensitivity for n-hexane.

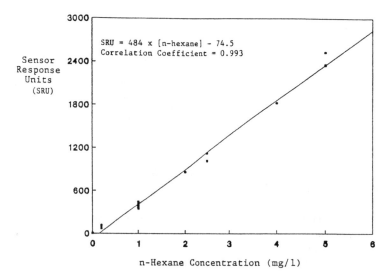

Figure 2. Linearity of the response of the Hydrocarbon Sensor when exposed to various levels of hexane.

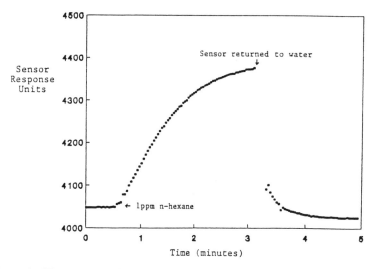

Figure 3. Hydrocarbon Sensor response to an injection of 1.0 ppm hexane.

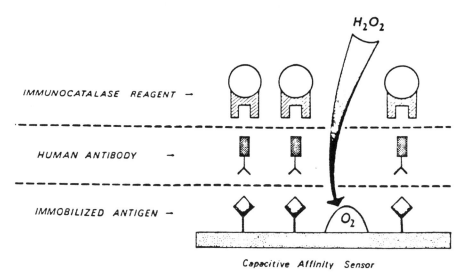

Figure 4. Capacitive Affinity Sensor for antibody detection.

then exposed to a diluted blood plasma sample. If the sample contained antibody directed against the antigen on the sensor, the antibody bound. Excess and irrelevant antibody were washed away. Next, an immunocatalase reagent was added to the sensor. The immunocatalase reagent consists of an antibody-binding protein conjugated to catalase. If antibody was bound to the sensor, then the immunocatalase conjugate likewise bound. Excess conjugate was washed away, and the sensor was exposed to H_2O_2 in solution. Oxygen bubbles were generated rapidly wherever catalase was present. The bubbles induced a change in capacitance which was monitored by the CASIS and recorded on a personal computer. Figure 5 shows the results of a sensor experiment in which two sensors were coated with p24, a recombinant protein from the HIV. Sensor #1 (A) was exposed to blood plasma from an individual with HIV infection. Sensor #2 (B) was exposed to plasma from an uninfected individual. The "1" on the graph indicates the time at which the H_2O_2 was added. Sensor #1, exposed to the HIV antibody-positive plasma, immediately responded due to the rapid generation of oxygen bubbles at its surface. Sensor #2, exposed to antibody-negative plasma, did not change. The results became apparent within seconds of adding H_2O_2.

Since each sensor response is independent of the others, an array of sensors can be used to test for antibodies to several antigens simultaneously. We have developed a multisensor array consisting of five sensors, each coated with a different HIV antigen or control substance. Figures 6 and 7 show the results of two experiments. In each experiment the sensors were coated as follows: Sensor #1 (A) was coated with a recombinant peptide from the gp41 envelope glycoprotein of HIV-1. (The gp41 peptide sequence is proprietary to BSC.) Sensor #2 (B) was coated with a recombinant protein version of p24, the major core protein of HIV-1. Sensor #3 (C) was coated with recombinant HIV-1 Protease. Sensor #4 (D) was coated with antibody against human immunoglobulin, and served as a positive control. Sensor #5 (E) was coated with an irrelevant human serum protein, and served as a negative control. Figure 6 shows the results of an experiment in which the plasma sample came from an individual without HIV infection; the only sensor to respond was #4, the positive control. In Figure 7, the plasma used was from an individual with HIV infection; all of the sensors responded in this experiment except #5, the negative control. Each of the experiments shown were the result of two five minute incubations: five minutes with antibody and five minutes with immunocatalase conjugate. The data were obtained in a total of about 12 minutes beginning with the addition of plasma to the coated sensors in the array. The total volume of plasma required for the five sensor array was 25 microliters.

The system described above can identify antibodies to three of the component proteins of the HIV. The multi-sensor array also provides two important internal controls to validate the results of the other sensors in the array. The information provided by the biosensor system is similar to that obtained by Western Blot in that it can detect patient antibodies specifically directed to the different proteins of the AIDS virus. It does so in much less

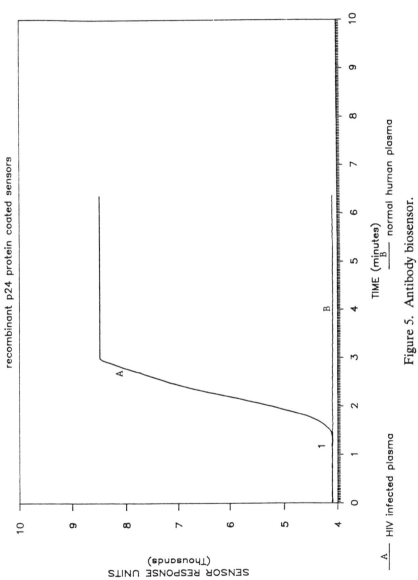

Figure 5. Antibody biosensor.

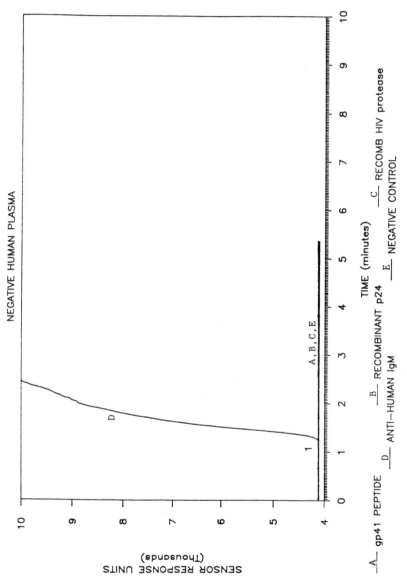

Figure 6. Multianalyte antibody biosensor.

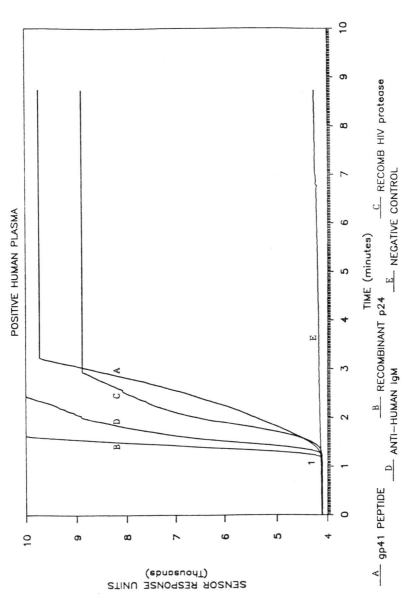

Figure 7. Multianalyte antibody biosensor.

time   than current screening tests, and requires only a minute quantity of blood.

Sensitivity and Specificity.   Table 2 shows the biosensor   results for 30 sera tested for antibodies to HIV gp41.   Sera were   obtained from 15 HIV-infected individuals and 15 individuals without   antibodies to HIV as determined by Western Blot.   The results show   sensitivity, specificity and accuracy all 100%. Results were obtained using five minute incubations each for antibody and immunocatalase conjugate. Total test time per sample was less than 15 minutes.

Automation of the Bubble Sensor.   Throughout the development of CASIS in all of its manifestations, manufacturability and ease of use have been considered essential to future development.   If a   device cannot be manufactured by conventional methods it will never be commercially viable. Likewise, the instrument must be easy to use and reasonably user-insensitive to be marketable.   The CASIS fits these characteristics.   A prototype device has been designed that can perform all necessary functions once the drop of blood has been added to a sample cassette.   Fluid handling, incubation timing and washing, as well as electronic monitoring and reporting of sensor responses are all handled by the device.   Sample cassette design allows for safe disposal of biohazardous materials once the test has been completed.   All components of the cassette and assay   device are manufacturable by well established automated processes.   New assays could be added to the system simply by changing the   coatings used in the cassettes.

Detection and Quantitation of Glucose

The bubble sensor was adapted to the detection of glucose by using a lectin with affinity for glucose (figure 8).   The appropriate lectin was attached to the amino-terminated silane on the sensor surface.   The sensor was then exposed   to   a   sample   either   containing   glucose   or   not. Glucosamine-conjugated catalase was then   incubated on the sensor.   The amount of catalase conjugate bound to the sensor, and therefore the rate of bubble formation, was inversely proportional to the amount of glucose bound. Figure 9   shows the results of an experiment in which eight sensors were simultaneously exposed to different concentrations of glucose.   If the sample lacked glucose then ample catalase conjugate was bound,   resulting in the fastest bubble-formation rate (sensors #1 and 2).   If the sample contained high amounts of glucose then virtually no catalase conjugate bound, resulting in no bubble formation   (sensors #7 and 8).   Intermediate concentrations yielded   intermediate rates of bubble formation.   Figure 10 shows the rate of bubble formation (Sensor Response Units per second) versus the   glucose concentration (ug/ml) in the sample.   The glucose   concentration of an unknown sample can be determined from this   standard curve. [Note: The range of glucose concentrations covered   by the standard curve shown in

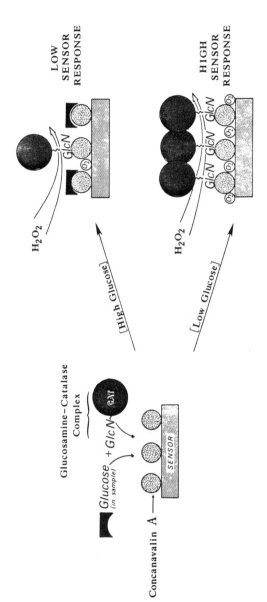

Figure 8. Capacitive Affinity Sensor for glucose.

TABLE 2

BIOSENSOR TESTS OF 30 HUMAN SERA FOR ANTIBODIES TO HIV gp41

| | | BIOSENSOR RESULTS | | |
		positive	negative	total
WESTERN	positive	15	0	15
BLOT	negative	0	15	15
RESULTS	total	15	15	30
		SENSITIVITY 15/15 = 100%	SPECIFICITY 15/15 = 100%	ACCURACY 30/30 = 100%

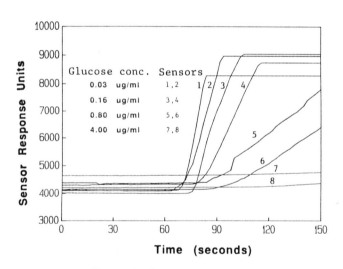

Figure 9. Glucose biosensor.

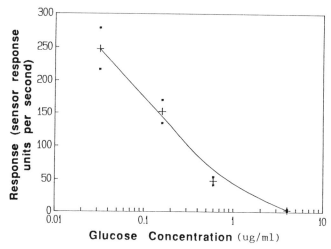

Figure 10. Standard curve.

Figure 10 is approximately 1/1000th of normal blood concentration. This corresponds to the concentrations of glucose found by transdermal collection of glucose from rats (data not shown).]

The Future of Capacitive Affinity Biosensors: Direct Detection of Antibodies

The CAS structure currently in use has been shown to be capable of detecting the binding of large molecules to its surface <u>directly</u>, that is, without the use of amplifying enzymes. The magnitude of the signal, however, is lower than that from the generation of bubbles. Routine use of direct detection would make the CASIS easier to use and to manufacture. Washing steps could be eliminated, simplifying ancillary machinery and fluid handling. Current research at Biotronic Systems Corporation is focused on accurately modeling the electric field and equivalent circuit of the CAS to optimize the detection of biomolecules directly. The next generation of Capacitive Affinity Biosensors will be more sensitive to make the detection of the binding event occurring on its surface more reproducible. The direct detection of biomolecules will open up vast new areas of use for the CAS. Direct detection of the binding event will make it possible to determine rates of binding in real-time, greatly simplifying affinity determinations. Any receptor-ligand interaction could be quickly examined in detail, including analysis of binding competitors.

Summary

The generic nature of the Capacitive Affinity Sensor has allowed the development of several methods within the context of this system. The hydrocarbon sensor is an excellent example of the capacitance principle at work; it is direct, rapid, sensitive, accurate, specific and inexpensive. The bubble sensor has been developed to detect several different analytes, particularly antibodies. The immunocatalase reagent binds to any mammalian antibody, permitting the current system to be used for detection of any antibody for which there exists a well characterized antigen. By using antibody-antigen competition, the current system can be used to detect any antigen for which there is a well characterized antibody. Lectins have been used effectively to detect the binding of sugars. More promising is the direct detection of large molecules which will allow the development of new tests for both diagnostic and analytical applications. Present research at Biotronic Systems Corporation is dedicated to the development of a direct detecting biosensor that performs sensitively and specifically, with reversibility comparable to that achieved in the hydrocarbon system. The CASIS presented here reveals only the first generation of a new and powerful analytical system.

Acknowledgments

The authors gratefully acknowledge the contributions of Ms. Sandra Toon for her recent work on biosensor assays. This work was supported in part by NIH SBIR Phase II Grant R44-AI26019. The following positive control reagent was obtained through the AIDS Research and Reference Reagent Program, Division of AIDS, NIAID, NIH: Human HIV Immune Globulin from Dr. Alfred Prince. The human sera used in the experiments described above, and Western Blot results for those sera were generously provided to BSC through a collaboration with Dr. Chyang Fang of the American Red Cross (Rockville, MD).

RECEIVED May 27, 1992

Chapter 10

Characterization of Fluorescent Dyes for Optical Immunosensors Based on Fluorescence Energy Transfer

Ai-Ping Wei, James N. Herron[1], and Douglas A. Christensen

Department of Pharmaceutics, University of Utah,
Salt Lake City, UT 84112

Eight fluorescent dyes of the fluorescein family, including fluorescein (F), tetramehtylrhodamine (T), rhodamine B (RB), lissamine rhodamine (LR), eosin (E), x-rhodamine (X), Texas Red (TR) and naphthofluorescein (NF), were labeled or colabeled onto mouse immunoglobulins to study their fluorescence energy transfer properties in solution. Of ten combinational pairs studied, the F-T and E-TR pairs were found to best satisfy a set of general criteria for biosensor applications. Fluorescence energy transfer properties of dye-IgG conjugates on hydrophobic surfaces were also studied using the total internal reflection fluorescence (TIRF) technique. Higher transfer efficiency was observed for the adsorbed molecules, which was attributed to the increased interactions between adsorbed molecules and possible conformational changes on surfaces. Multivariate methods, i.e., multiple linear regression and factor analysis, were used to further characterize the donor-acceptor interactions.

Often described as a spectroscopic ruler at the molecular level (1), fluorescence energy transfer is the transmission of excited state energy from a donor molecule to an acceptor molecule. The process is radiationless and is primarily the result of dipole-dipole interactions between the donor and acceptor. Typical of dipole-dipole interactions, the efficiency of energy transfer decreases with the sixth power of the distance between the donor and acceptor molecules. Most donor-acceptor pairs have characteristic distances of about 50 Å (2). For this reason, researchers were able to achieve various homogeneous immunoassays by means of fluorescence energy transfer (3, 4). As a typical example, antigens and antibodies are labeled with fluorescent donor and acceptor molecules, respectively. In the absence of sample, the labeled antigen binds to

[1]Corresponding author

0097–6156/92/0511–0105$06.00/0

the antibody and fluorescence energy transfer occurs with high efficiency between the donor and acceptor molecule. This results in a majority of the fluorescence emission coming from the acceptor species. Upon the addition of sample, the labeled antigens are displaced by analyte antigens. As acceptor molecules move away from donor molecules, the efficiency of energy transfer decreases and the fluorescence emission is due almost entirely to the donor species. Homogeneous immunossays can be achieved by monitoring relative changes in the fluorescence intensities at emission maxima of the donor and acceptor. Therefore, fluorescence energy transfer is ideal for the biosensor applications because it offers significant advantages in terms of simplicity as compared with other assay methods such as enzyme immunoassays. In fact, this concept has recently been used in studies of chemical and immunosensors (5, 6).

The conditions necessary for energy transfer are (7): a) the fluorescence spectra of donor molecule must overlap with the absorption spectra of the acceptor molecule; b) the donor molecule must be a fluorescent group with sufficiently a lifetime of a few nanoseconds; c) the donor and acceptor must be within a certain distance (e.g.~50 Å). For biosensor applications, however, donor-acceptor dye pairs should satisfy additional criteria in order to develop sensors characterized by high sensitivity, low cost, and simple operation. For example, the absorption spectra of donor and acceptor should have very little overlap in order to minimize the direct excitation of the acceptor molecule and hence reduce background fluorescence. Also, the fluorescence maxima of both donor and acceptor should be at a wavelength higher than the fluorescence of serum which usually occurs around 500-515 nm (8). Third, donor and acceptor molecules should have high extinction coefficients and high fluorescence quantum yields to ensure high sensitivity. Fourth, excitation spectra of donors should match the spectral lines of inexpensive excitation sources, e.g., lasers or laser diodes. Although no particular donor-acceptor pair will completely satisfy all of these conditions, fluorescein and its analogs are believed to be especially well-suited for biosensor applications. This is because these compounds exhibit strong visible fluorescence at wavelength between 500 and 610 nm, and more than ten different analogs are commercially available in chemically-reactive forms. The focus of this paper was to evaluate the spectral properties of these dyes and their potential use as fluorescence energy transfer donors and acceptors. Eight dyes were examined in these studies; they included fluorescein, eosin, tetramethylrhodamine, rhodamine B, lissamine rhodamine B, X-rhodamine, Texas Red, and Naphthofluorescein. Issues investigated included: a) energy transfer properties in solution; b) energy transfer properties at interfaces; and c) multivariate analysis of the donor-acceptor interactions.

Materials and Methods

Preparation of Immunoglobulin-Dye Conjugates. Mouse immunoglobulins (Organon Teknika-Cappel, Malven, PA) were allowed to react with the amine-reactive forms of fluorescent dyes in 50 mM phosphate buffer, pH 7.7, for about 12 hours at room temperature. The structures of eight fluorescent dyes used in this study are given in Figure 1. Rhodamine B isocyocynate was the product of Research Organics (Cleveland, OH). Eosin-5-isothiocyanate, the sulfonyl chlorides of lissamine

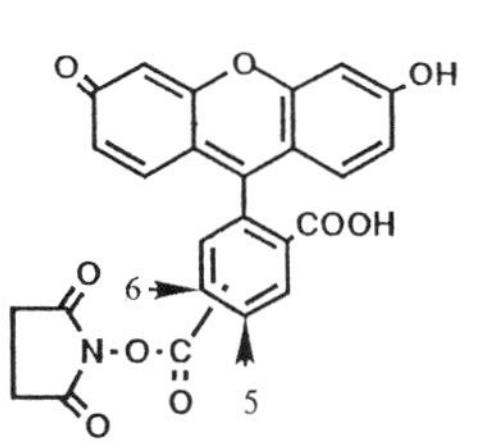

Figure 1. Eight fluorescent dyes that have been attached to mouse immunoglobulins to study their spectral properties and their potential use as fluorescence energy transfer donors and acceptors. The abbreviated names of these compounds are given in the parentheses.

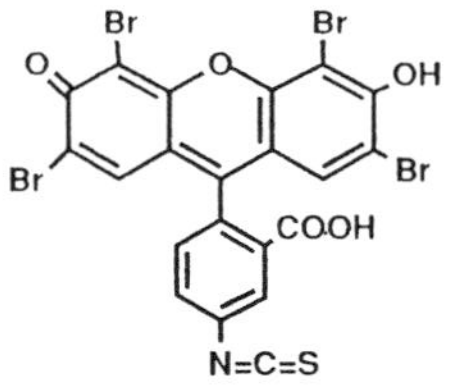

Eosin-5-isothiocyanate

(Eosin, E)

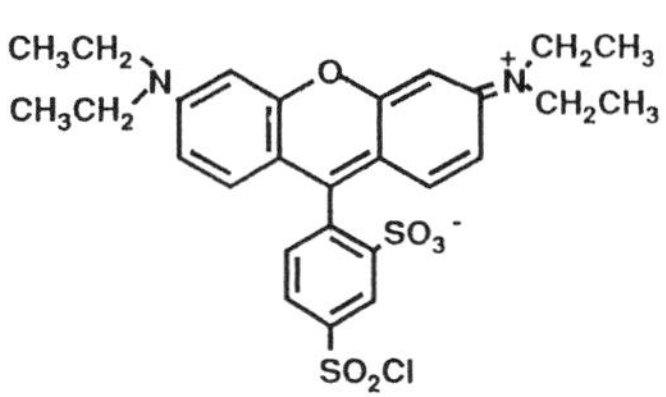

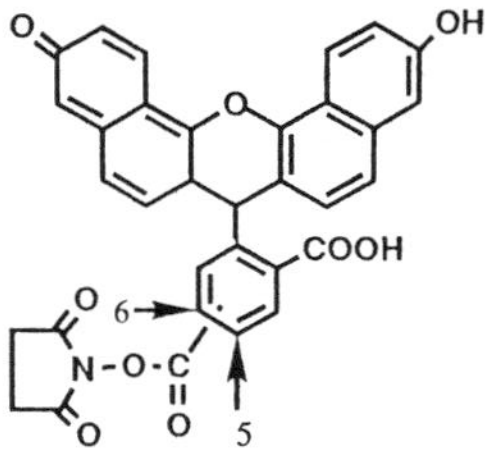

5-(and-6)-carboxynaphtho-fluorescein, succinimidyl ester

(Naphthofluorescein, NF)

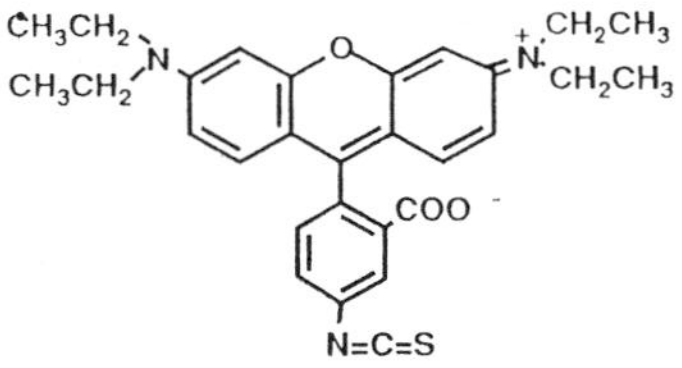

Rhodamine B-5-isothiocyanate

(Rhodamine B, RB)

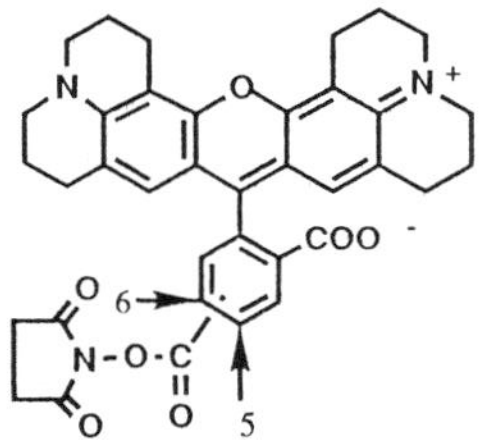

5-(and-6)-carboxy-X-rhodamine, succinimidyl ester

(X-rhodamine, X)

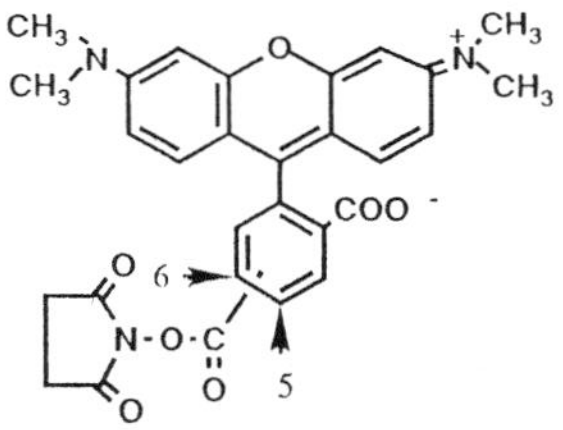

5-(and-6)-carboxyltetramethyl rhodamine, succinimyl ester

(T-rhodamine, T)

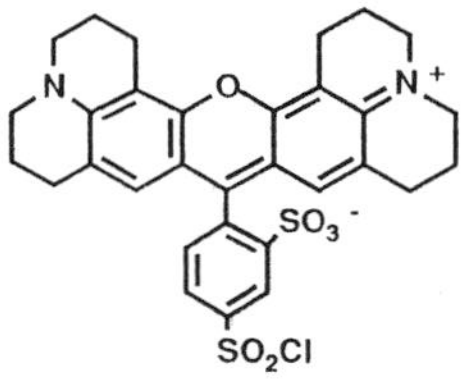

Texas Red sulfonyl chloride

(Texas Red, TR)

5-(and-6)-carboxylfluorescein, succinimdyl ester

(Fluorescein, F)

Lissamine Rhodamine B, sulfonyl chloride

(Lissamine, LR)

rhodamine B and Texas Red, the succinimidyl esters of 5-,6-carboxyfluorescein, 5-, 6-carboxytetramehyl rhodamine, 5-,6-carboxy-X-rhodamine and 5-,6-carboxynaphthofluorescein were all purchased from Molecular Probes, Inc. (Eugene, OR). Immunoglobulins colabeled with donor and acceptor fluorophors (D-A-IgG) were prepared by co-reacting an equimolar mixture of donor and acceptor dyes with IgG. Labeled proteins were separated from unreacted dyes on a PD-10 gel filtration chromatography column (Pharmacia LKB, Piscataway, NJ).

Absorption Spectra, Fluorescence Spectra and Fluorescence Lifetimes. A Perkin-Elmer Lambda 2 spectrometer was used for all the UV-Vis spectra measurements. Fluorescence spectra were measured with a Greg-200 multi-frequency phase fluorometer (ISS, Champaign, IL) in the photon counting mode. Fluorescence lifetimes were measured on the same fluorometer. Dyes were excited with the 488 nm and 514 nm lines of an Argon ion laser.

Adsorption on DDS Surfaces. Silica surfaces were derivatized with dimethyl chlorosilane (DDS) according to the following procedure: a) Quartz slides (*ca.* 2.4x4.8cm) were cleaned in hot chromic acid for 45 min., rinsed extensively with double deionized water, and dried at 100 °C oven; b) Clean slides were incubated with DDS (10%) and toluene (90%) for 30 min; c) The slides were then rinsed with ethanol and water; d) The slides were dried under N_2 at 100°C for 1 hour; e) The slides were stored in jar covered with aluminum foil for adsorption experiments.

The fluorescence of labeled immunoglobulins adsorbed to DDS surfaces was measured by total internal reflection fluorescence spectroscopy using a charged-coupled device (CCD) for optical detection. An argon-ion laser was used as the excitation source. Immunoglobulin samples were prepared in 100 mM phosphate buffer (pH7.7). The degree of labeling was between 2 and 4 dye molecules per IgG. In a typical experiment, a 0.3 mg/ml solution of the labeled IgG was added to the flow cell and allowed to adsorb for 45 minutes, the cell was then flushed with buffer and the spectrum was measured.

Multivariate Analysis. The composite absorption or fluorescence spectra of a donor-acceptor pair were resolved into its constituent components using the multiple linear regression model:

$$F_{a\text{-}d} = \alpha \cdot F_d + \beta \cdot F_a + \varepsilon \qquad (1)$$

where , F_d, F_a, $F_{a\text{-}d}$ are the absorption spectra or the fluorescence spectra of the D-IgG, A-IgG and D-A-IgG, respectively; α, β are the linear coefficients to be determined, and ε is the error term. The regression was done on an Apple Macintosh SE/30 using the software **StatWorks** (Cricket Software Inc., Philadelphia, PA). The error term was found to be under 2%.

The fluorescence spectra of immunoglobins labeled with fluorescein (F-IgG), rhodamine B (RB-IgG), and colabeled with both F and RB (F-RB-IgG) were subject to factor analysis in order to determine the interaction components between the donor and acceptor dyes. The UV-Visible absorption spectra and fluorescence spectra of the F-IgG, RB-IgG, F-RB-IgG conjugates and the mixture of F-IgG & RB-IgG were measured at several different concentrations. A 176x17 matrix was constructed from 17 emission spectra (F-IgG, RB-IgG, F-IgG & RB-IgG mixtures, and F-RB-IgG). Factor analysis was performed using the statistical software package **Number Cruncher Statistical System** (NCSS, Kaysville, Utah) on an IBM PC.

Results and Discussions

Energy Transfer Properties in Solution. Each of the eight dyes was conjugated to IgG and absorption and emission spectra were recorded. The wavelengths of maximum (λ_{max}) absorption and fluorescence of the dye molecules are given in parentheses in Table 1. Based on the spectral properties of each dye-IgG conjugate, ten donor-acceptor combinations were selected as potential energy transfer pairs. The relative merits of each pair were judged according to the criteria described in the introduction. The results of these studies are listed in Table 1. In brief, F-T and E-TR or E-X were identified as the best pairs according to the above criterion. The argon-ion laser has high intensity bands at 488 and 514 nm, the frequency-doubled Nd:YAG laser has a strong band at 532 nm, and the helium-neon laser has a band at 543 nm. Also, serum fluorescence has a maximum at about 500 nm, but does not produce significant interference at wavelength greater than 530 nm. Since the λ_{max} of F-IgG and E-IgG absorption are 496 and 527 nm, the F-T and E-TR pairs can be excited with a 488-nm argon-ion laser and a 532-nm Nd:YAG laser, respectively. The E-TR pair avoids serum interference and is ideal for both blood and urine samples. Although the F-T pair would not be optimal for blood samples, it should work for urine samples. Moreover, since there have been numerous energy transfer studies of the F-T combination (2, 3, 4, 7), it makes a good model system for biosensor research.

Some of the other pairs in Table 1 do have unique properties, although they do not satisfy all of the above criteria. For example, the fluorescence of fluorescein (F) was quenched significantly by rhodamine B (RB), but little increase in the fluorescence of RB was observed. The most plausible explanation for this observation is that energy transfer from fluorescein to rhodamine B is a "dark" process; i.e., energy is transferred from donor to acceptor, but the acceptor dissipates it through a non-radiative process. The T-X, T-TR and LR-TR pairs all exhibited excellent overlap between the emission spectrum of the donor and the absorption spectrum of the acceptor; however, the emission maxima of the donor and acceptor were not well-resolved. Spectral separation technique would be required in order to sort out the contributions of donor and acceptor.

In addition to spectral properties, fluorescence lifetime is another important parameter to characterize the energy transfer properties of donor and acceptor molecules. Since fluorescence energy transfer is a resonance process between two oscillating dipoles, sufficient lifetime of the donor molecule is a condition necessary for such a process to occur. Under otherwise similar conditions, donor molecules with longer lifetimes would result in higher energy transfer efficiency (9). Table 2

Table 1   Fluorescence Energy Transfer Properties of the Eight Dyes in Figure 1

	Pair I (Excitation at 488 nm)				Pair II (Excitation at 543 or 532 nm)					
Donor	Fluorescein (496, 519)	Fluorescein (496, 519)	Fluorescein (496, 519)	T-rhodamine (520/556, 582)	T-rhodamine (520/556, 582)	T-rhodamine (520/556, 582)	Lissamine (574, 590)	Lissamine (574, 590)	Eosin (527, 545)	Eosin (527, 545)
Acceptor	T-rhodamine (520/556, 582)	X-rhodamine (584, 605)	Rhodamine B (559, 585)	X-rhodamine (584, 605)	Texas Red (598, 610)	N-Fluorescein (607, 670)	Texas Red (598, 610)	N-Fluorescein (607, 670)	Texas Red (598, 610)	X-rhodamine (584, 605)
Advantage	The F-T is an excellent pair by all the criterion	Absorption and fluorescence spectra are well separated.	Very good spectral overlap; high transfer efficiency.	Excellent spectra overlap; long emission wavelength	Good spectral overlap; and long emission wavelength	T-NF pair has longest emission wavelength	Good spectral overlap; long emission wavelength	Good spectral overlap; long emission wavelength	E-TR is a good pair by most of the criterion.	E-X is a good pair by most of the criterion.
Dis-Advantage	None	The energy transfer efficiency will be low.	The F-RB can result in dark transfer.	Absorption and fluorescence spectra are not well separated.	Absorption and fluorescence spectra are not well separated.	NF will show significant fluorescence only when pH > 9.5.	Spectra were not well separated.	NF will show significant fluorescence only when pH > 9.5.	Eosin has the shortest life-time among the dyes.	The thiourea bonds from isothiocyanates are not stable.

summarizes the lifetime values obtained for six dye-IgG conjugates. Eosin had the shortest lifetime of 1.6 ns and exhibited a two-step decay. This was attributed to a large and highly allowed S_o - S_1 transition and also to the heavy atom quenching by bromine atoms (10). The other five dyes all showed a single-step decay. Noteworthy are Texas Red and X-rhodamine, both of which have longest lifetimes, 4.5 ns and 4.8 ns, respectively, and are known to have longest fluorescence emission wavelength (λ_{max}~610 nm) and highest extinction coefficients (ε_{max}=8-10 x10^4 $M^{-1}cm^{-1}$) among the dyes studied. Selection of these dyes as energy transfer acceptors for biosensors will help to avoid interference by serum fluorescence and to increase sensitivity.

Energy Transfer Properties at Interfaces. Total Internal Reflection Fluorescence (TIRF) Spectroscopy is especially well-suited for measuring fluorescence intensity and spectra of adsorbed molecules (8). As far as optical detection is concerned, TIRF is probably the simplest and yet the best model system for investigating the application of evanescent excitation and florescence energy transfer for sensor development. The fluorescence spectra of E-TR-IgG and F-RB-IgG in adsorbed states are shown in Figure 2 and Figure 3, respectively. For E-TR-IgG, the the intensity ratio, F_{610}/F_{545}, increased from 0.5 in bulk solution (spectra not shown) to 0.83 for the adsorbed conjugates, while that of the F-RB-IgG conjugates F_{585}/F_{545} changed from 0.5 in bulk to 1.5 when adsorbed. The E-TR pair showed lower transfer efficiency than the F-RB pair because the degree of spectral overlap between donor fluorescence and acceptor absorption is less for the former than the later. Other donor-acceptor-IgG conjugates also exhibited higher energy transfer in the adsorbed state than in bulk solution. Because the energy transfer efficiency is proportional to R^{-6}, these results are probably due to the close packing of IgG molecules at the interface, which brings dye molecules on adjacent IgG molecules close enough for energy transfer. Also, denaturation of IgG molecules at the interface may further contribute to the increase in energy transfer. To investigate these alternatives, the experiments were repeated under the same conditions except that the donor and acceptor fluorophors were attached to different IgG molecules instead of being colabeled to the same IgG molecule. Specifically, two mixtures of F-IgG and T-IgG were prepared such that the concentration of T-IgG was held fixed while the F-IgG concentration differed. Absorption and fluorescence spectra of the two samples in bulk solution are given in Figure 4 and Figure 5, respectively. The slight shoulders in the fluorescence spectra probably resulted from direct excitation of T-IgG molecules. Energy transfer was virtually non-existent in these mixtures because the donor and acceptors were not within energy transfer distance in such mixtures. However, when they were adsorbed onto the DDS surface, the dyes were brought much closer than when they were in solution due to the tight packing of IgG at the interface. As a result, the fluorescence of T-rhodamine increased dramatically, as shown in Figure 6. Although the T-IgG concentration was fixed, the ratio of intensities at 520 nm and 580 nm increased from F_{520}/F_{580}=0.24 to F_{520}/F_{580}=0.46 as the F-IgG concentration varied from A_{494}=0.0214 to A_{494}=0.0913. This experiment clearly indicated that energy transfer efficiency changed with variations in donor concentration. In a competitive fluoroimmunoassay, the presence of analyte will alter the concentration of an antigen labeled with a donor fluorophor in the vicinity of an antibody labeled with an acceptor

Table 2. Life-times of Six Dye-IGG Conjugates in 100 mM Phosphate Buffer (pH7.7) (Phase±0.5°, MOD±0.01)

	τ_1 (ns)	$\tau_1\%$	τ_2 (ns)	$\tau_2\%$	Chi^2
F-IGG	3.381	98.3%	0	1.7%	6.3
E-IGG	1.632	77.7%	0.279	22.3%	1.1
T-IGG	2.492	96.7%	0	3.3%	0.9
LR-IGG	1.856	94.7%	0	5.3%	2.9
TR-IGG	4.466	96.9%	0	3.1%	12.2
X-IGG	4.77	97.9%	0	2.1%	19.2

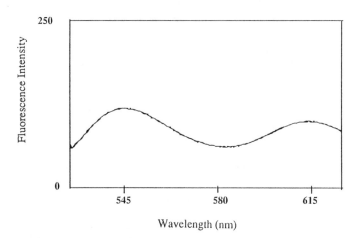

Figure 2. The fluorescence spectra of E-TR-IgG adsorbed on DDS surfaces, measured after 45 minutes of adsorption when the bulk proteins was flushed away with buffer solution. The intensity ratio F_{610}/F_{545} changed from 0.5 in bulk to 0.83 for the adsorbed states, indicating enhanced energy transfer.

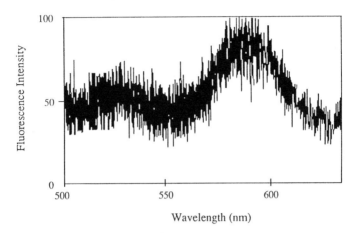

Figure 3. The fluorescence spectra of F-RB-IgG adsorbed on DDS surfaces. The intensity ratio F_{585}/F_{520} changed from 0.4 in bulk to 1.5 when adsorbed. These spectra are noisy because the F-RB combination is a 'dark' transfer pair.

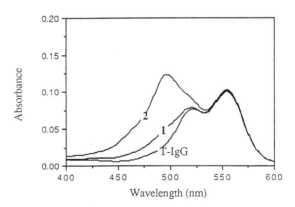

Figure 4. The absorption spectra of two mixtures of F-IgG and T-IgG. The concentration of T-IgG was fixed at $A_{555} = 0.1016$, while the F-IgG varies from $A_{494} = 0.0214$ to $A_{494} = 0.0913$.

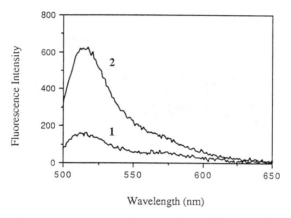

Figure 5. The fluorescence spectra of the samples in Figure 4. The shoulder at 580 nm is due to direct excitation of the rhodamine. Energy transfer is virtually non-existent because the donors and acceptors are not close enough in these mixtures.

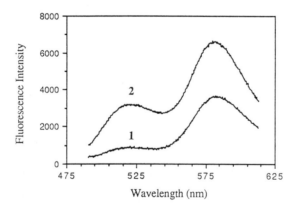

Figure 6. The fluorescence spectra of the two samples adsorbed on DDS surfaces. The change in the relative intensity at 520 nm and 580 nm suggests that energy transfer between the F-IgG and T-IgG has increased significantly in the adsorbed states, as compared to the bulk spectra in Figure 5.

fluorophor. This will result in a decrease in energy transfer efficiency. Since immunochemical reaction should have much better control over distances than simple mixtures of donors and acceptors, improved results are expected when antibodies and antigens are actually involved.

Multivariate Analysis of the Fluorescence Spectra of the Energy Transfer Donor and Acceptors. Multivariate analysis is a relatively new but rapidly expanding approach to data analysis (11). Various multivariate techniques have been used in spectroscopic analysis for calibration (12), background correction (13), and extracting components from mass spectra of mixtures (14). These techniques, however, have not been specifically noted in the literature to be applied to analyze spectra of fluorescence energy transfer donors and acceptors. As will be noted in the following discussions, multivariate techniques have proven to be useful mathematical tools for spectra separation, background correction, and corrections for trivial reabsorption.

Multiple Linear Regression. Energy transfer efficiency can be calculated from the quenching of donor fluorescence according to:

$$E = \frac{F_d/A_d - F_d^a/A_d^a}{F_d/A_d} \qquad (2)$$

where, E is energy transfer efficiency; F_d is the fluorescence intensity of donor in the absence of acceptor; A_d is the absorbance of donor in the absence of acceptor; F_d^a is the fluorescence intensity of donor in the presence of acceptor; A_d^a is the absorbance of donor in the presence of acceptor. In order to determine E accurately, both the fluorescence and absorption spectra of the D-A-IgG complex should be separated into components of D-IgG and A-IgG. A multiple linear regression model (Equation 1) was used for the spectral separation. An example of this method is shown in Figure 7 for the fluorescein-rhodamine B pair. In this figure, the measured spectrum of F-RB-IgG is compared to spectra of the two individual components and their sum. Both the t-test and F-test suggested highly significant results at a significance level of less than 0.0001. The error term was < 1%, and the coefficient of determination was > 0.99. These all indicate that Equation 1 produced an excellent fit of the experimental data and that the result is highly significant. Similarly, the composite UV-vis absorption spectra of dye-IgG conjugates were separated into their constituent components. Energy transfer properties of this particular pair can be evaluated from the quenching of donor and fluorescence and from the enhancement of acceptor fluorescence (Figure 8). As described above, the F-RB pair seems to be an example of dark transfer because even though 94% fluorescence of the fluorescein was transferred to rhodamine B, there was little increase in rhodamine B fluorescence.

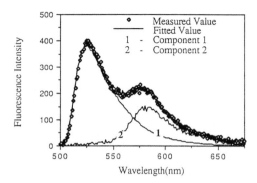

Figure 7. The fluorescence spectra of the F-RB-IgG conjugates was separated into its constituent components by a multiple linear regression method described in the text. The good overlap between measured and the fitted spectra indicates that the model was appropriate.

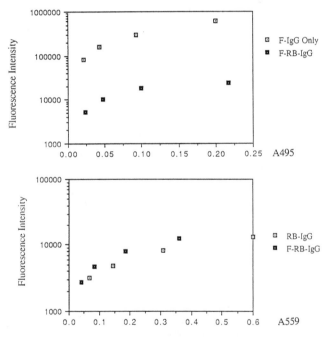

Figure 8. Energy transfer properties of fluorescein (donor) and rhodamine B (acceptor). The upper plot shows the quenching in fluorescence intensity of the F-IgG in the presence of RB as function of absorbance at 495 nm. The lower plot shows the increase in fluorescence intensity of the RB-IgG in the presence of F as function of absorbance at 559 nm. The energy transfer efficiency calculated in terms of quenching F-IgG is 94%, while the fluorescence of rhodamine showed little increase. This indicates that the F-RB is a dark transfer pair.

Factor Analysis. Residual values from the above the multiple linear regressions are plotted as a function wavelength in Figure 9. It is interesting to note that systematic patterns exist at high concentrations and tend to disappear at lower concentrations. According to basic principles of linear regression, an ideal fit should result in evenly-distributed residuals. When systematic patterns exist in the residual plot, there are usually interactions between regressing components (15). In order to characterize the interaction term between the donor and acceptor at high concentrations, the spectral data were subject to multivariate factor analysis. A 176x17 matrix was constructed from 17 emission spectra (F-IgG, RB-IgG, F-IgG & RB-IgG mixture, and F-RB-IgG) at several concentrations. Four significant factors were identified by the factor analysis (Figure 10). The four factors accounted for 99.62% of the original variations. The first two factors (components) were the emission spectra of fluorescein and rhodamine B, respectively, and accounted for 98.62% of the original variations. The third factor varied with the intensity of the signal, and was attributed to electrical noise in the detection system. The fourth factor was correlated to the absorption spectrum of rhodamine B and was due to the interaction between donor (F) and acceptor (RB) through trivial reabsorption.

Once these factors are identified, they can be used in several different ways. For example, analytical functions of these factors can be incorporated into the regression model of Equation 1 as additional components so that the regression coefficients (α and β) can be determined more accurately. This practice is named Principal Component Regression (PCR) and has been used widely for background correction in spectroscopic analysis (13). The interaction term due to trivial reabsorption can also be corrected once its contribution to the original spectra is known. Although this term is normally small, it may become significant for molecules adsorbed at interfaces due to the increase in local concentration. This is especially important for biosensors that involve immunochemical reactions at the surfaces of solid substrates.

Conclusions

In summary, we have studied the spectral properties of eight different fluorescent dyes from the fluorescein family in the form of dye-IgG conjugates. The absorption maxima of the dyes span from 490 nm to 600 nm, and emission maxima from 520 to 610 nm. Of the eight dyes, ten combinational donor-acceptor pairs were evaluated for their energy transfer properties in the form of donor-acceptor-IgG conjugates. The merits of each pair were judged by consideration of their potential applications in simple and homogeneous biosensor apparatus with high sensitivity for both blood and urine samples. The F-T and E-TR pairs were found to be especially well-suited for this application. We have also studied the fluorescent energy transfer properties of the donor-acceptor pairs at the interface between aqueous solution and DDS surfaces. Higher energy transfer was observed at the interface for all the pairs studied. This was attributed to the decrease in intermolecular distance and possible conformational changes or rearrangement of the adsorbed IgG molecules. Two multivariate techniques were applied for the first time to analyze spectroscopic data of fluorescence energy transfer. A multiple regression procedure was successfully applied to separate the

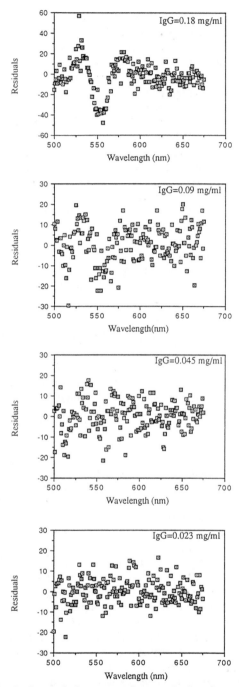

Figure 9. Residual plots of the regression analysis of F-RB-IgG conjugates (F/IgG=3.6, RB/IgG=3.4). The systematic changes at higher concentrations tend to disappear as concentration decreases, suggesting that some additional interaction between donor and acceptor exists at higher concentrations.

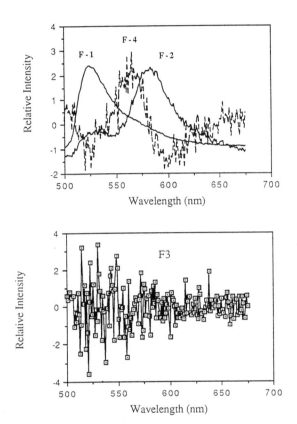

Figure 10. Results of factor analysis with a 176x17 data matrix constructed from spectra of F-IgG, RB-IgG, F-RB-IgG and F-IgG/RB-IgG mixtures at several concentrations. The relative intensities are normalized values. The four significant factors resulting from this analysis account for 99.62% of the original variation. The first two components (F-1 and F-2 in the upper plot) correspond to fluorescein and rhodamine B, respectively. The third component (F-3 in the lower plot) is a factor related to light scattering and PMT noises. The fourth component (F-4 in the upper plot) is an interaction term between F and RB.

complex spectra of a donor-acceptor pair into its constituent components. Factor analysis was able to identify four major components from the fluorescence spectra of the F-RB-IgG conjugates. The first two components corresponded to donor and acceptor spectra. The third component corresponded to background noise, and the fourth one accounted for the trivial reabsorption term. The combination of these two techniques is believed to be a powerful mathematical tool for the development of accurate and sensitive biosensors.

Acknowledgments

The authors would greatly appreciate the funding of this work by AKZO Corporate Research America, Inc. Also greatly acknowledged are J.-N. Lin, V. Hlady and J. D. Andrade for their helpful discussions and assistance.

References

1. Stryer, L. *A. Rev. Biochem.* **1978**, 11, 203.
2. Khanna, P. L.; Ullman, E. F., *Anal. Biochem.* **1980**, 108, 156.
3. Ullman, E. F.; Khanna, P. L., *Methods in Enzymol.* **1981**, 74, 28.
4. Khanna, P. L. in *Nonisotopic Immunoassay;* Ngo, T. T., Ed.; Plenum Press, New York, NY, **1988**; 211.
5. Barnard, S. M.; Walt, D. R. *Science*, **1991**, 251, 927.
6. Anderson, F. P.; Miller, W. G. *Clin. Chem.*, **1988**, 34, 1417.
7. *Fluorescence Spectroscopy, an introductiuon for biology and medicine;* Pesce, A. J.; Rosen, C. G.; Pasby, T. C., Eds., Marcel-Dekker, Inc., New York, 1971.
8. Wolfbeis, O. S.; Leiner, M. *Anal. Chim. Acta,* **1985**, 167, 203.
9. *Principles of Fluorescence Spectroscopy;* Lakowicz, J. R., Ed.; Plenum Press, New York, NY.
10. Chen, R. F.; Scott, C. H. *Analytical Letters.* **1985**, 18 (A4), 393.
11. *Multivariate Data Analysis;* Hair, J. F.; Tatham, R. L., Eds.; McMillan, 1987.
12. Beebe, K. R.; Kowalski, B. R. *Anal. Chem.*, **1987**, 59, 1007 A.
13. Gemperline, P. J.; Boyette, S. E.; Tyndall, K. *Appl. Spectr.* **1987**, 41, 454.
14. Windig, W; Meuzellar, H. L. C. in *Computer-enhanced Analytical Spectroscopy,* Meuzelaar, H. L. C.; Isenhour, T. L., Eds. Plenum Press, 1987, 67.
15. *Applied Linear Regression Models;* Neter, J; Wasserman, W.; Kutner, M. H. Eds. 2nd edition, Irwin, Boston, MA, 1987.

RECEIVED May 12, 1992

Chapter 11

Concanavalin A and Polysaccharide on Gold Surfaces

Study Using Surface Plasmon Resonance Techniques

R. F. DeBono[1], U. J. Krull[1], and Gh. Rounaghi[2]

[1]Chemical Sensors Group, Erindale Campus, University of Toronto, Mississauga, Ontario L5L 1C6, Canada
[2]Department of Chemistry, Faculty of Science, Mashhad University, Mashhad, Iran

Dynamic studies involving deposition and subsequent denaturation of selective binding proteins, or of complexes formed by selective interactions, provide insight into problems of reversibility and useful lifetimes of biosensors. The adsorption of concanavalin A (Con A) in the presence and absence of Ca^{2+} and Mn^{2+} onto polycrystalline gold surfaces and subsequent interactions of the protein with the polysaccharides glycogen and dextran have been investigated *in situ* using surface plasmon resonance (SPR) spectroscopy and surface plasmon microscopy (SPM). The adsorption of concanavalin A and subsequent reactions of the protein with the polysaccharides were followed by observing changes of reflectivity at a fixed angle of incidence as a function of time. Our results suggest that Con A exhibits non-Langmuirian adsorption kinetics. The polysaccharide binding involves two Langmuirian processes based on adhesion and subsequent dehydration of the polysaccharide layer leading to compaction. The negligible effect of ethylenediaminetetraacetic acid (EDTA) on the reversal of the complexation reaction between dextran and Con A at the surface provides further evidence of the dense structure of this layer. Characterization by SPM shows these surfaces to be smooth and homogeneous on the 4 μm scale except for the Au/glycogen/Con A system which was allowed to react for periods greater than 10 hrs.

0097–6156/92/0511–0121$06.00/0

Recent interest in the areas of biocompatibility and development of biosensors has prompted studies of the adsorption or covalent deposition of proteins and polymers onto solid surfaces (1-4). Changes in the conformation or orientation of proteins or protein-ligand complexes at solid/liquid interfaces during and after the adsorption process may occur and can result in the denaturation of the protein or complexes (5,6). Such processes can determine the useful lifetime and reversibility of biosensors.   Applications involving the close contact of biological materials with a foreign surface must account for surface induced changes. The optical techniques of surface plasmon resonance (SPR) spectroscopy and microscopy (SPM) can be used for the physical characterization and real time observation of dynamic events at an organic film located in close proximity to a metal surface.

One system that is representative of a selective interaction is the adsorption of concanavalin A (Con A) to a polycrystalline gold surface, and the subsequent selective interaction of the protein with a polysaccharide. Con A is a well characterized lectin, found in jack bean, which reacts to form a precipitate with a restricted group of branched polysaccharides (α-D-mannopyranosyl) in a manner analogous to an antibody-antigen reaction (8). At pH 7.0 Con A exists as a tetramer(if ionic strength > 0.3) with a net negative charge (9). The isoelectric point is approximately 5. At low ionic strength (<0.3) or at a pH of 5.0 Con A exists primarily as a dimer (10). Each subunit is a globular protein which contains 237 amino acids (MW 25,500) and a single saccharide binding site (11,12).   The monomer is an elliptical prolate with a cross-section of 4.0 x 3.9 nm and a length of 4.2 nm. The molecular surface is smooth with a large cavity that extends deep into the molecule. Ellipsoidal dimers are formed by pairing across an axis of two-fold symmetry forming units that are 8.4 x 4.0 X 4.0 nm in size. The dimers are paired back-to-back to form tetramers (13).

The saccharide binding site is activated by divalent cations (optimal for Ca^{2+} and Mn^{2+} ions), which induce a conformational change in each subunit (14,15).   Con A can bind non-specifically to cell surfaces, hydrophobic materials (16) and lipid vesicles (17) through a hydrophobic cavity which is independent of the saccharide binding cavity (18). Con A will generally bind to saccharide chain ends containing terminal non-reducing α-D-glucopyranosyl moieties (10).

Surface Plasmon Spectroscopy and Microscopy

Figure 1 shows a prism of refractive index n_1 coated with a thin metal layer with a complex refractive index n_m in contact with a medium of refractive index $n_o < n_1$. Parallel polarized light of wavelength λ passing through the prism at an angle of incidence greater than the critical angle (that is defined by the prism and the outer medium) reflects at the metal/prism boundary. A reflectivity curve (SPR curve) is produced (Figure 1b) when the angle of incidence is varied while monitoring the reflected intensity. If the metal layer is sufficiently thin (<100 nm) then the reflectivity curve will exhibit a minimum at an angle Θ_{min}. This minimum arises as a result of resonant excitation involving the collective motion of conduction band electrons (surface plasmons) with a coherence length of L_c on the order of 10 µm for gold. Associated with this oscillating charge density wave is a strong evanescent wave with a penetration depth (d_p) into the outer contacting medium (n_o) of approximately 300 nm for $\lambda = 632.8$ nm. If an organic film of refractive index n_f is place between the metal and the outer medium the reflectivity curve will be modified owing to its strong dependence on the environment near the metal surface.

Maxwells' equations indicate that if the refractive index of this layer $n_f > n_o$ then the SPR curve will shift to a larger angle of incidence as the thickness of the film (d_f) increases, and the angle becomes constant when $d_f > d_p$.

The surface plasmon resonance condition is determined by the spatial average of the thickness and refractive index of the organic film as sampled over the coherence length (L_c) near the metal surface.

Differences in the profile of the film over distances greater than L_c will be detected as spatially resolved differences in reflected intensity at a fixed angle of incidence. This forms the basis of surface plasmon microscopy (19,20) which uses these differences in intensity to provide optical contrast to image regions of different optical mass (refractive index, thickness). Further details concerning the theory of surface plasmon resonance can be found in a review by Raether (7).

Experimental Section

Reagents: The following reagents were obtained commercially: highly purified concanavalin A type IV (Sigma Chemical Co.): dextran-T500 (Pharmacia Fine Chemicals); Glycogen (Oyster,0.25M NaCl on dry material, BDH biochemicals); ethylene diaminetetracetic acid (Fisher Sci. Co.); calcium chloride $CaCl_2.2H_2O$ (BDH Chemicals); Manganese chloride $MnCl_2.4H_2O$ (Sigma Chemical Co.). All of the chemicals employed in this study were of analytical reagent grade and were used without further purification. All distilled water was obtained using a Milli-Q five stage cartridge purification system (Millipore, Mississauga, Ontario) and was degassed under vacuum before use. Premium quality glass microscope slides were obtained from Fisher Scientific (Pittsburgh, PA). Gold with a purity of greater than 99.99% was obtained from DEAK International and chromium rods were purchased from R.D. Mathis Co. (Long Beach, CA).

Equipment: Metal films were prepared using a Key High Vacuum Products model KV-301 metal vapour deposition instrument (Nesconset, New York). The thicknesses of organic films in air were determined using an Auto EL II, nulling reflection ellipsometer (Rudolph Research, Flanders, NJ). The optical source was a 1 mW continuous wave helium-neon laser (632.8 nm) at fixed angle of incidence of 70.00°. Data was analyzed using a "Film 85" software package, version 3.0, program #70 to determine NS and KS of substrate and program #13 to determine the thickness of the organic layer assuming a refractive index of 1.5. NS and KS for each bare gold substrate was determined before starting each experiment.

Static contact angle (wettability) estimates were done by imaging a 10 µl drop of distilled water with a monocular microscope which contained a reticle protractor.

Surface Plasmon Spectrophotometer and Microscope: The experimental apparatus used for SPR and SPM is shown in Figure 2. Angular control was achieved by the use of a goniometer from a Rudolph ellipsometer in order to vary the angle of incidence with a precision and accuracy of 0.005 degrees.

The optical train consisted of a 2 mW linearly polarized helium-neon laser (Spectra Physics, Hughes) which emitted at 632.8 nm with a beam diameter of 2 mm, a divergence of 0.01 mradians and a stability better than 0.8 %; a dichroic polarizer (Melles Griot) set at 45° and two Glann Thompson polarizers (referred to as Pol 1 and Pol 2). The optical beam was refracted at the first prism face and underwent total internal reflection at the prism base. The beam was then directed by a mirror to pass through an attenuation filter and a diffuser (scotch tape) onto a Hamamatsu R-928 photomultiplier tube (PMT).

The mirror was removed and a 5x microscope lens was positioned so that the focal point was at the base of the prism in order to do surface plasmon microscopy. The scattered light was collected by the lens to produce an image which was captured by a Hitachi KP-111 CCD solid state camera (Nedco, Mississauga, ONT.). The automatic gain control of the CCD camera was disabled.

For SPR work Pol 1 was set at 45 degrees while Pol 2 was set at either 0.00° (parallel) or 90.00° (perpendicular). SPM work required careful control of the intensity of the source which could be achieved by changing the polarization of Pol 1.

Procedures

Preparation of Glass Wafers: Glass slides were cleaned by soaking in a detergent solution in an ultrasonic bath for one hour at 60 °C followed by extensive washing in water and drying in an oven for several hours. Wafers were then transferred to the vacuum deposition system where a thin layer of chromium (1.5-3.0 nm) was deposited at 0.05 nm.sec^{-1} at 5 µTorr. The bell jar was then back filled with prepurified argon and the chromium rod was exchanged for a molybdenum boat containing gold. During this step the chromium layer was exposed for a few minutes to atmospheric conditions resulting in the formation of thin chromium oxide layer. The system was then evacuated to 5 µTorr and approximately 45.0 nm of gold was deposited at a

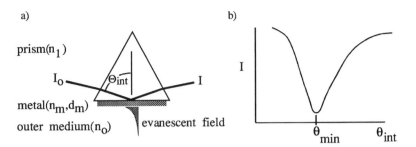

a)

prism(n_1)

I_0

Θ_{int}

I

metal(n_m,d_m)

outer medium(n_0)

evanescent field

b)

I

θ_{min} θ_{int}

Figure 1: a) Arrangement for optical excitation of surface plasmons; b) Reflected intensity versus angle of incidence for a surface plasmon resonance curve.

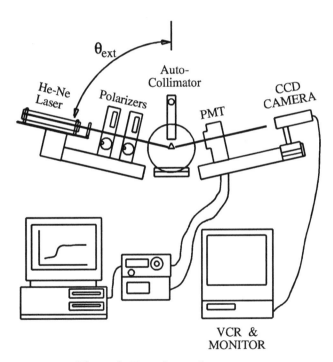

θ_{ext}

Auto-Collimator

He-Ne Laser

Polarizers

PMT

CCD CAMERA

VCR & MONITOR

Figure 2: Experimental setup

rate of 0.4 nm.sec⁻¹.   Wafers were transferred to plastic slide cases and wrapped in parafilm until needed.

Prism Assembly: A 60.00° BK-7 prism (30.00 x 30.00 x 30.00 mm) was optically connected to the gold coated glass wafer using a matching immersion oil (type B, Gargille Laboratories Inc., n = 1.515). The prism was then clamped into a teflon flow-through cell (150 ul) and a viton O-ring was pressed against the glass slide as shown in Figure 3. A Microperspex peristaltic pump (LKB Bromma) was used to deliver solutions to the cell.

The entire assembly was transferred to the sample stage of the SPR apparatus and an autocollimator was used on the portion of the wafer extending past the prism to ensure that the plane of incidence was normal to the glass slide. A manual reflectivity scan was carried out, recording the reflected intensity from 70.00° to 90.00° in 0.50° steps for both parallel and perpendicular polarization modes, and the minimum angle of incidence was determined. From this scan a fixed angle of incidence was chosen for the time dependent measurements. The angle was chosen to be approximately 1 degree less than the minimum angle, where a compromise between sensitivity and signal-to-noise ratio was achieved. The intensity of the source was recorded at the beginning and end of any time based measurements to verify that drift in laser intensity had not occurred.

At the end of the experiment a second SPR curve was obtained to ensure that the system was behaving properly. Imaging was then carried out. Note all angles of incidence reported in this paper are internal angles of incidence.

Determination of Normalization Factor: All intensity measurements were converted to R_p/R_s reflectivity measurements by multiplying the ratio of I_p/I_s (Parallel intensity/Perpendicular intensity) by a normalizing factor NF. The normalization factor corrected for reflectivity losses at the prism faces, mirror and attenuation filter, and for differential response of the PMT due to movement of the optical beam across the PMT surface as the angle of incidence was varied. The NF was obtained from a reflectivity scan using a blank wafer which was optically connected to the prism, such that $NF = I_s'/I_p'$.

Modelling and Fitting Program: SPR data was modelled and analyzed using a program developed by R. DeBono in Quick Basic 4.5 (SPMOD2). The model was based on the use of the exact Fresnel equations to calculate R_p/R_s for a N layered system, where N could be varied (21). An iterative fitting program varied the initial specified parameters by carrying out a one-dimensional minimization along favourable directions in N-dimensional space using Powell's method (21). The best fit was established by use of a χ^2 parameter. Error estimates in the fitted values were obtained by varying the specified value until the χ^2 doubled.

Results and Discussion

Surface plasmon resonance spectroscopy and microscopy were used to investigate the adsorption of concanavalin A onto polycrystalline gold surfaces, and to observe subsequent interaction of the protein with the polysaccharides dextran and glycogen.

The reflectivity curve for a 44.6 nm thick gold film in contact with water (Figure 4) has a moderately sharp and well defined minimum. Hence gold is capable of supporting surface plasmons and is expected to show sensitivity to optical mass changes occurring at the surface of the metal.

Furthermore, the full width at half height associated with the reflectivity curve corresponds to a coherence length of 4 μm for the surface plasmon defining the lateral resolution of this surface for SPM.

The polycrystalline gold surfaces had a wettability angle of 60 ± 5° indicating a partially hydrophobic surface. X-ray photoelectron spectroscopy indicated a thick gold film with no chromium present at the surface.

Figure 4 shows the resonance curves for three systems: a bare gold wafer in contact with water, after exposure to 85 nM Con A in the presence of 100 μM Ca^{2+} and Mn^{2+}, and after subsequent exposure to 1.46 μM glycogen at a flow rate of 3.5 ml.min^{-1}. Figure 5 shows the dynamic changes in reflectivity at a fixed angle of incidence of 70.483° for the above additions.

The positive shifts of θ_{min} in the resonance curves are indicative of the addition of an organic layer with a refractive index larger than that of the contacting medium (n = 1.3330). Optical properties (refractive index and thickness) of the metal were obtained from the bare gold system. The theoretical response function of the bare Au system based on the Fresnel equations was determined for an organic layer of refractive index 1.5000. Shifts in the experimentally determined minimum angle and reflectivity could then be converted to average thickness values. These results are summarized in Table I and Figures 6 and 7.

Table I. Observed Minimum Angle Shift (ΔΘ) and Corresponding Average Layer Thickness as a Result of Exposure of the Gold Surface in the (I) Presence of 100 μ Ca^{2+} and Mn^{2+} to 85 nM Con A and Subsequent Exposure to 1.46 μM Glycogen

System	Ions Present	Solutions added	Δθ ± 0.006	d (nm) ± 0.03
I	$Ca^{2+,}$ Mn^{2+}	85 nM Con A	0.597	2.81
		1.46 uM Glycogen	0.736	3.32

Adsorption of Con A: On exposure of the gold surface to Con A a rapid adsorption process was observed over the first 53 seconds (Figure 6). Flushing the surface with water resulted in only a slight decrease (Figure 5) in reflectivity probably due to removal of weakly physisorbed Con A.

The rate of adsorption of material to a surface is a function of the diffusion of molecules to the surface and subsequent binding affinity. The diffusion rate will be a function of the concentration and flow rate while the

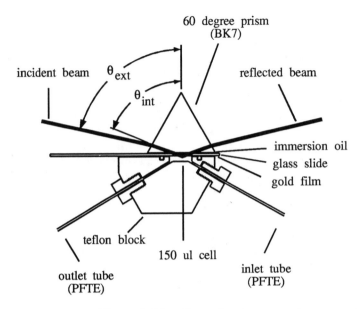

Figure 3: Flow-through setup

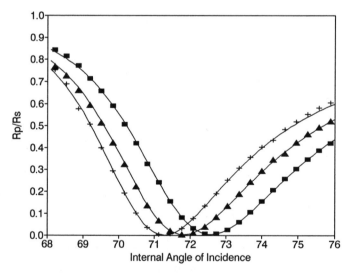

Figure 4: Surface plasmon resonance curves for a (+) bare gold wafer in contact with water, (▲) after exposure to 85 nM Con A in the presence of 100 μM Ca^{2+}, Mn^{2+} and (■) subsequent addition of 1.46 μM glycogen. Solid lines are the theoretical curves corresponding to a smooth homogenous film of refractive index 1.5 with a thickness of 0, 2.52 and 5.98 nm. Gold: ε_m = -12.54, 1.35, d_m = 44.6 nm. Contacting medium: n = 1.3330.

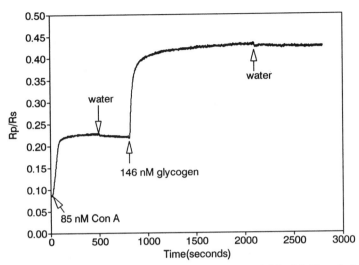

Figure 5: Response (θ = 70.483°) for the addition of 85 nM Con A in the presence of 100 μM Ca^{2+}, Mn^{2+} and subsequent addition of 1.46 μM glycogen.

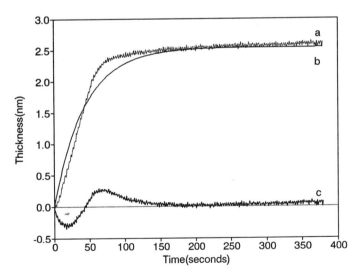

Figure 6: a) Response curve (θ = 70.483°) for the addition of 85 nM Con A in the presence of 100 μM Ca^{2+}, Mn^{2+} b) Langmuirian adsorption model (solid black line), B = 30 nm, K = 0.02 sec^{-1}. c) difference between experimental and Langmuirian curves.

binding affinity is a function of the number of available sites and interaction energies. The design of the flow cell provided a thin relatively unstirred layer of solution in contact with the metal and resulted in diffusion control of protein translocation across this layer to the surface. The relatively irreversible binding of Con A to the surface established an effective solution concentration at the surface which was very small and unchanged with respect to surface occupation by the protein.

The adsorption kinetics for Con A in the presence of Ca^{2+} and Mn^{2+} do not follow a simple Langmuirian model as can be seen by comparison of the theoretical curve and experimental curve in Figure 6. Initially (first 53 sec) adsorption of Con A appears to be diffusion controlled with a constant rate of adsorption such that the average thickness increases by 0.040 nm.sec^{-1} until a critical breakpoint thickness of 19.7 nm is reached. An initial diffusion limited adsorption step is consistent with the initial linear rate of protein deposition observed for all concentrations of Con A shown in Figure 8, and with the dependence of Con A adsorption on the solution flow rate. The initial adsorption process was rapid and irreversible in contrast with the diffusion process. Subsequent adsorption of the protein then becomes a site limited kinetic process which was best fitted as a double Langmuirian model

$$T = B_1 [1 - \exp(-K_1 t)] + B_2 [1 - \exp(-K_2 t)]$$

where T is the average thickness and t is the time (B_1 = 0.41 ± 0.03 nm,

K_1 = 0.082 ±0.006 sec^{-1}; B_2 = 0.26 ± 0.03 nm, K_2 = 0.009 ± 0.0003; Norm χ^2 = 0.46). This suggests the presence of two distinct binding site populations.

The solution conditions used in this work suggest that Con A existed predominantly in the dimer form. Crystallized Con A consists of compact ellipsoidal domes of dimensions 4.2 x 4.0 x 4.0 nm while the dimer has dimensions of 8.4 x 4.0 x 4.0 nm (23).

The adsorbed Con A layer reached a limiting average thickness of 2.81 ± 0.03 nm which did not correspond to any of the previous dimensions, suggesting that Con A had denatured upon adsorption (rapidly with respect to diffusion rates).

Figure 8 shows the adsorption of Con A at a flow rate of 2.6 ml.min^{-1} for concentrations of the protein ranging from a high of 78 nM to a low of 3.9 nM in the presence of 100 µM Ca^{2+} and Mn^{2+}. Note that the actual concentrations in solution may be lower than those stated due to adsorption of Con A to the walls of the volumetric flask used to prepare these solutions; relative error will be the largest at the lowest concentrations. Adsorption behaviour similar to that observed in Figure 6 was observed for all concentrations of Con A, but the limiting thickness was concentration dependent as shown in table II.

At high concentrations Con A is expected to adsorb in a purely random fashion with little chance given for each individual molecule to attain a minimum energy configuration at the surface (denaturation) before another

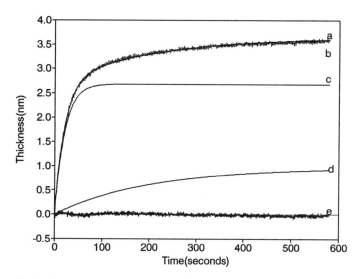

Figure 7: a) Response curve ($\theta = 70.483°$) for the addition of 1.46 μM glycogen b) combined double Langmuirian model (solid black lines), normalized $\chi^2 = 0.23$ c) Langmuirian model component $B_1 = 2.68 \pm 0.03$ nm, $K_1 = 0.0492 \pm 0.005$ sec^{-1} d) Langmuirian model component $B_2 = 0.97 \pm 0.05$ nm, $K_2 = 0.0050 \pm 0.0006$ sec^{-1}. e) difference between experimental and fitted double Langmuirian curves.

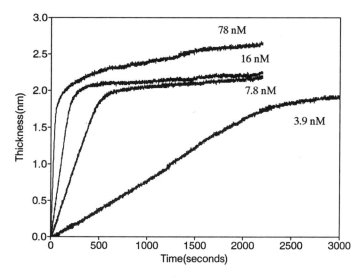

Figure 8: Response curves ($\theta = 70.483°$) for the addition of 3.9, 7.8, 16 and 78 nM Con A in the presence of 100 μM Ca^{2+}, Mn^{2+}

molecule is adsorbed nearby and sterically prevents further spatial denaturation. At low concentrations the adsorption rate is sufficiently slow to allow the protein to attain a minimum energy configuration, which may lead to significantly ordered structures at the surface which are more denatured (reduced thickness).

Table II. Observed Minimum Angle Shift ($\Delta\Theta$) and Corresponding Average Layer Thickness as a Result of Exposure of the Gold Surface in the Presence of 100 μM Ca^{2+} and Mn^{2+} to 78 nM Con A

[Con A] (nM)	$\Delta\Theta$ ($\pm$ 0.006)	d ($\pm$ 0.03 nm)	Ellipsometry (nm)
78	0.585	2.74	2.3 $\pm$ .3
16	0.489	2.32	1.9 $\pm$ .1
7.8	0.489	2.34	1.5 $\pm$ .3
3.9	0.457	2.13	1.3 $\pm$ .4

For all the curves in Figure 8, a similar critical thickness was associated with the point at which a decrease in the adsorption rate was observed. This break point could be associated with a surface coverage at which the adsorption process begins to become site limited, or at which specific sites evolve.

Wafers were removed from solution, dried under argon, and then studied by ellipsometry. The same trend in thickness values was observed and the smaller values of thickness reported by ellipsometry suggest that the removal of water during the drying step results in further compaction of the protein layer.

Selective Interactions of Con A and Polysaccharide: Addition of glycogen in the presence of Ca^{2+} and Mn^{2+} to the adsorbed layer of Con A results in an increase in thickness of 3.5 nm (Figure 7).   This adsorption process is fitted best as a double Langmuirian curve with rate constants of 0.05 sec^{-1} and 0.005 sec^{-1} and prefactors of 2.68 and 0.97 nm respectively.   This may be due to a fast initial adsorption of highly extended and hydrated polysaccharide which then slowly collapses with the expulsion of water.   To examine if such a compaction would result in a net increase in reflectivity at a fixed incident angle we considered the following approximation.   Glycogen was expected to adsorb in a highly extended and hydrated configuration with a refractive index of 1.3457 corresponding to a 10% sucrose solution and an average thickness of 16.5 nm.   We estimated that the layer compacted by a factor of 5 to a thickness of 3.3 nm corresponding to the 50% sucrose system with a refractive index of 1.4654.   Modelling this using the exact Fresnel equations gave values for R_p/R_s at 70.483° of 0.1714 and 0.2467 for the uncompacted and compacted systems respectively. Hence the SPR technique is capable of responding to conformational changes of the polysaccharide at the surface as suggested.

Table III. Observed Minimum Angle Shift ($\Delta\Theta$) and Associated Thickness in the (II) Presence or (III) Absence of 100 μ Ca^{2+} and Mn^{2+} as a Result of Exposure of the Gold Surface to 40 nM Con A and Subsequest Exposure to 120 nM Dextran and (IV) for Exposure to 125 nM Dextran Followed by 75 nM Con A, Then 125 nM Dextran and Final Exposure to 100 μM EDTA

System	Ions Present	Solutions Added	$\Delta\theta$ ±.006	d(nm) ±0.03
II	$Ca^{2+,}$ Mn^{2+}	40 nM Con A	0.507	2.43
		120 nM Dextran	0.109	0.50
III		40 nM Con A	0.450	2.19
		120 nM Dextran	0.000	0.00
IV	$Ca^{2+,}$ Mn^{2+}	125 nM Dextran	0.187	0.90
		75 nM Con A	0.381	1.85
		125 nM Dextran	0.109	0.50
		100 uM EDTA	0.000	0.00

System II of Table III consists of the adsorption of Con A to the gold surface and subsequent adsorption of the polysaccharide dextran. Similar behaviour is observed as in the Con A/glycogen system but the thickness of the dextran layer is much smaller (0.50 ± 0.03 nm) compared to that of glycogen (3.32 ± 0.03 nm).

In system III Con A was adsorbed to the gold surface in the absence of Ca^{2+} and Mn^{2+}. Subsequent exposure to dextran resulted in no change in the reflectivity or shift in the minimum angle. This confirmed that Con A when adsorbed to the gold surface in the presence of Ca^{2+} and Mn^{2+} must be at least partially active for selective saccharide binding to occur and that the dextran was not simply physisorbed to the surface.

In system IV, dextran was initially physisorbed to the gold surface and achieved a limiting thickness of 0.90 ± 0.03 nm. This provided confirmation when taken with the results from system III that the gold surface was completely covered by the protein in the previous systems. Metallic gold provided a surface for the adsorption of dextran, and this adsorption of dextran was not observed when (system III) unactivated Con A was first adsorbed to the gold surface. In system IV activated Con A was then adsorbed to the dextran layer, resulting in an increase of thickness of 1.81 nm and similar adsorption kinetics as observed previously (i.e. non-Langmuirian). Exposure of this layer to more dextran resulted in a further increase in thickness by a value of 0.50 nm.

To examine if the last system (gold/dextran/Con A/dextran) was close packed, a 100 μM EDTA solution was pumped past the surface for 60 hrs to remove Ca^{2+} and Mn^{2+} via complexation (24), which would deactivate the polysaccharide binding site in Con A. After 60 hrs no shift in the minimum angle or change in reflectivity was observed. This suggests that the metal ions are unavailable to the EDTA due to steric blockage by the adsorbed dextran layer.

Surface Plasmon Microscopy

Surface plasmon microscopy was used to search for microscopic structure within the systems studied. No structure was observed for any of the gold, gold/Con A or gold/Con A/dextran surfaces at the 4 μm resolution level, suggesting that nucleation and growth of the organic layers was occurring at the sub-microscopic level and that these films were optically homogeneous (allowing the use of the Fresnel equations).

Structure was observed only for the gold/glycogen/Con A systems after incubation times of 10 hrs. Figures 9a and 9b provide representative images of this system for angles of incidence of 72.53° and 73.68° respectively. Image 9b is the inverted image of 9a, i.e. bright regions in 9a correspond to dark regions in 9b indicating that two distinct layers of different thickness exist. The thickness between these layers based on the incident angle difference of 1.15° required to invert the optical contrast corresponds to a thickness difference of approximately 4.5 nm. Hence the bright regions in Figure 9a correspond to regions which are 4.5 nm thicker than the dark regions. The polarization when switched from parallel to perpendicular resulted in the loss of the image confirming that this was a truly an image based on SPR.

Conclusion

The SPR and SPM experiments indicated that Con A adsorbed to the gold surfaces initially by a diffusion limited process until a critical coverage was achieved. Subsequent adsorption of protein then followed a double Langmuirian model suggesting the existence of two distinct binding site populations. Growth and formation of domains of protein were not observed at the microscopic level and suggest relative homogeneity of the protein layer. Some control of distribution of the spatial organization of the protein film could be achieved by varying the concentration of the protein, where high concentrations provided kinetic conditions which reduced denaturation. Complexation of Con A with polysaccharide was shown to be a selective process based on activation of the saccharide binding site. Reversal of the complexation by deactivation of the binding site by removal of the divalent metal ions was not possible due to the time-dependent compaction of the bound polysaccharide overlayer which blocked the connection between the protein layer and bulk solution. This experimental result may demonstrate a generic physical phenomenon that may be detrimental to the long-term function of some antibody-based biosensors, particularly at higher antigen concentrations.

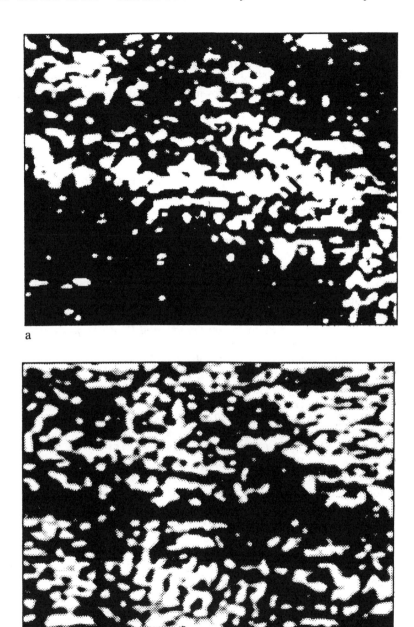

a

b

Figure 9: Enhanced SPM images of Gold/glycogen/Con A system after a reaction time of 14 hours. Region imaged is 400 μm by 200 μm. a) θ = 72.53°; b) θ = 73.68°

Acknowledgements

The authors would like to thank the Natural Sciences and Engineering Research Council of Canada for financial support of this work. We thank J.D. Brennan and R.S. Brown for useful discussions.

Literature Cited

1. Liedberg, B.;Ivarsson, B;Lundstrom, I. J. *Biochem. BioPhys. Methods* **1984**,9, 233.
2. Liedberg, B.;Lundstrom, I.J.; Wu. C. R.; Salaneck, W.R. *J. Colloid Interface Sci* **1985**, 108, 123.
3. Slaneck, W. R.; Lunstrom, I.; Liedberg, B; *Prog. Colloid and Poly. Sci* **1985**, 70, 83.
4. Jonsson, U.; Malmquist, M.; Ronnbeg, I.; Berghem, L. *Prog. Colloid and Poly. Sci.*, **1985**, 70, 96.
5. Soderquist, M.E.; Walton, A.G. *J. Colloid Interface Sci.* **1980**, 75, 386.
6. McMillin, C.R.; Walton, A.G. *J. Colloid Interface Sci.* **1974**, 48, 345
7 (a)Raether, H. *Surface Plasmon Oscillations and Their Applications.* Physics of Thin Films, Vol. 9, Academic Press, New York, **1977**, 145.
   (b)Raether, H. *Surface Plasmons on Smooth and Rough Surfaces and on Gratings*; Springer, Berlin, **1988**
8. Summer, J.B; Howell, S.F. *J. Bacteriol.* **1936**, 32, 227
9. McKenzie, G.H.; Sawyer, W.H.; **1973**, 248, 549
10. Schnebli, H.P.; Bittiger, H., Eds; *Concanavalin A as a Tool* John Wiley and Sons, New York
11. Yariv, J.; Kalb, A.J.; Levitzki A. *Biochem. Biophys. Acta.* **1968**, 165,303
12. So, L.L.; Goldstein, I.J. *J. Biol. Chem.* **1968**, 165, 303
13. So., L.L.; Goldstein, I.J *J. Biolog. Chem.* **1967** 242 1617
14. Harrington, P.C.; Wilkins, R.G. *Biochemistry*, **1978**, 17, 4245
15. Sherry, A.D.; Buck, B.E.; Peterson, C.A. *Biochemistry* **1978**,17, 2169
16. Ochoa, J.L.; Kristiansen, T.; Pahlman, S. *Biochem. Biophys. Acta* **1979**, 577, 102
17. Edelman, G.M.; Wang, J.L. *J. Biol. Chem.* **1978**, 353, 3016
18. Reeker, G.N. Jr.; Beker, J.W; Edelman, G.M. *J. Biol. Chem.* **1975**, 250, 1525
19. Rothenhausler, B.; Knoll, W., *Nature* **1988**, 332, 6115
20. Hickel, W.; Knoll, W.; *Thin Solid Films* **1990**, 187, 349
21. Heaven, O.S.; *Optical Properties of Thin Films*, Dover **1965**
22. Press, W.H.; Flannery, B.P.; Teukolsky, S.A.; Vetterling, W.T; *Numerical Recipes*, Cambridge University Press, New York, **1988**, 294
23. Becker, J.W; Reeke, G.N; Wang, J.L.; Cunningham, B.A; Edelman, G.M.; *J. Biol. Chem.* **1975**, 250(4), 1513
24. Brown, R.D.; Brewer, C.F.; Koenig, S.H. *Biochemistry* **1977** 16, 3883

RECEIVED April 3, 1992

Chapter 12

Noninvasive Determination of Moisture and Oil Content of Wheat-Flour Cookies

Near-Infrared Spectroscopy in the Wavelength Range 700–1100 nm

R. M. Ozanich, Jr., M. L. Schrattenholzer[1], and J. B. Callis

Department of Chemistry, Center for Process Analytical Chemistry, University of Washington, Seattle, WA 98195

Near-infrared spectroscopy in the wavelength range 700-1100 nm was evaluated in transmission and reflectance modes for the analysis of moisture and oil in wheat flour based cookies *in situ*. Spectral data were pretreated by means of normalization, which significantly decreased multiplicative effects arising from the high degree of light scatter. It was found that the wavelength of 965 nm (a combination band of H_2O) correlated best to moisture, while the wavelength of 930 nm (the third overtone CH stretch of methylene) correlated best to oil. These wavelengths corresponded to the maximum absorbance of water and oil respectively. Computer simulations facilitated interpretation of experimental results. Relative errors for moisture determination were 3.8% for the analysis of multiple samples. Relative errors for oil measurement were 6-7.5% for a large variety of cookies with varying morphologies and physical characteristics. Analytical precision of transmission and reflectance sampling methodologies were approximately equal. However, reflectance measurements were found to be biased towards the surface, and therefore transmission measurements may be preferred where uniform sampling is desired.

Moisture and oil content are key parameters for routine assessment of product quality of baked goods. Current methods for evaluation of these parameters are time consuming and costly to carry out and moreover must be performed off-line. This is unfortunate because, if such information were available

[1]Current address: Walla Walla College, Walla Walla, WA 99362

0097–6156/92/0511–0137$08.00/0
© 1992 American Chemical Society

on-line, inferior product could be detected immediately and the process conditions altered so as to minimize loss and maintain quality consistently at the highest level.   Methods of moisture determination in baked goods range from simple oven drying (gravimetry) to complicated procedures such as the Karl Fischer titration and digestion/extraction techniques (1, 2, 3). Analysis times vary from hours to days for the various procedures. Similarly, the methods for oil determination are lengthy and complex and include digestions, extractions, and filtrations (1, 2, 3). With the exception of gravimetry, these methodologies   require separation of the analyte prior to analysis with consequent destruction of the sample. However, even gravimetry irreversibly alters the sample.

Near-Infrared Spectroscopy

An attractive alternative to these methods is near infrared (NIR) spectroscopy in the wavelength range 1100-2500 nm. Overtone and combination bands in the NIR derive from fundamental vibrational absorptions in the mid-infrared. Excitation of a molecule from the ground vibrational state ($v=0$) to a higher vibrational state ($v>1$) results in overtone absorptions.   Combination absorption bands arise when two different molecular vibrations are excited simultaneously. Due to their forbidden nature (i.e., the low frequency of occurrence of transitions for which $\Delta v > 1$), NIR bands are several orders of magnitude less intense than the fundamental bands in the IR. However, the relatively low extinction coefficients in the NIR yield good linearity of absorbance with analyte concentration and permit long, convenient pathlengths to be used. Absorptions in the NIR are mainly due to C-H, O-H, and N-H bond stretching and bending motions. Therefore, NIR is well-suited for the analysis of organic compounds.

Typically, near-infrared spectroscopy (NIRS) is performed in the wavelength range 1100-2500 nm, where it has found numerous applications in agriculture and food industry (4, 5). One of the first, and certainly the most well-developed NIR technique, is the simultaneous determination of protein and moisture in wheat (6).   But further research has lead to assays for constituents such as fiber, starch, and total carbohydrates in a variety of agricultural products and processed foods (7).

An essential ingredient in this technology is a diffuse reflectance geometry. Generally, the material to be analyzed is reduced to a uniform powder as the sole preparation step. Low absorbance cross-sections of these powders permit linear correlation of the inverse log of the diffusely reflected intensity with analyte concentration. However, due to the broad overlapping bands, multivariate statistical calibration techniques must be used in order to extract meaningful analytical information (8).   One unique advantage of multivariate calibration is its ability to extract "non-chemical" properties such as the baking quality of flour from the spectra (9).

Because of its ability to make measurements in highly scattering samples, NIRS has seen application to on-line process analysis. For example, it has been evaluated for the determination of bran in a flour-milling operation *(10)*, and on-line measurements of oil, starch, and gluten in a wet-corn mill solid cake *(11)*.

However, certain disadvantages and limitations of the technique of near-infrared reflectance analysis would appear to prevent its universal applicability to process situations. Conventional NIR diffuse reflectance spectroscopy is not well-suited to the analysis of thick inhomogeneous objects. Generally, the grinding of a sample to a uniform, fine particle size is an essential preparation step. One further disadvantage is the shallow depth of penetration (0.1-1.0 mm) of NIR radiation in the wavelength region of 1100-2500 nm *(5)*. Thus, it would appear that NIRS could not be successfully applied to on-line analysis of such inhomogeneous objects as cookies, crackers, and other baked goods.

Short-Wavelength NIR. There exists an under-appreciated region of the spectrum in the range 700-1100 nm, known as short-wavelength near-infrared (SW-NIR), which offers numerous advantages for on-line and *in-situ* analyses. This portion of the NIR region is accessible to low-cost, high performance silicon detectors and fiber-optics *(12, 13)*. NIR bands in this region are noticeably weaker than the first and second overtones, and this leads to good linearity of log inverse diffuse reflectance versus analyte concentration for a wide variety of irregular objects such as seeds, nuts *(14)*, and polymer pellets *(15)* presented without any sample preparation. The depth of penetration of SW-NIR radiation is also much greater, permitting a more adequate sampling of the bulk material. A number of applications of this spectral region to constituent analysis have been successfully accomplished *(12, 16, 17, 18)*.

Experimental

The instrumental apparatus used for these studies is shown in Figure 1. The spectrophotometer was a Pacific Scientific Co. (Presently NIRSystems, Silver Spring, MD) Model 6250 Near-Infrared Spectrophotometer. Both reflectance and transmission sampling geometries were investigated. Reflectance sampling was accomplished using a bifurcated fiber-optic probe containing two concentric rings of fibers (Figure 2A). The outer ring of fibers delivers the light to the sample and the inner circle of fibers collects the light from the sample, transferring it to the silicon detector. Transmission sampling was performed using separate fiber-optic illumination and detection probes (Figure 2B). The two fiber-optic probes were mounted a fixed distance apart, typically 1 cm, and the sample was placed between them.

Sample temperature was not controlled for these studies. All samples were analyzed at room temperature (20-25 °C), which did not vary by more than 2-3 °C over the course of an experiment. Although the spectral influence of small temperature changes is subtle, improved analytical performance is expected for temperature controlled studies. A high purity alumina ceramic disk (thickness ca. 2.5 mm) was used as a reference for reflectance sampling *(19)*, while an air reference was used for transmission sampling schemes. To

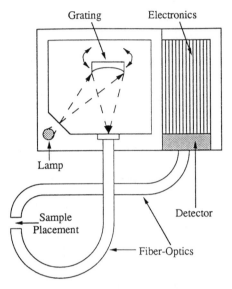

Figure 1. Instrumental apparatus shown with fiber-optics for measurement in transillumination mode.

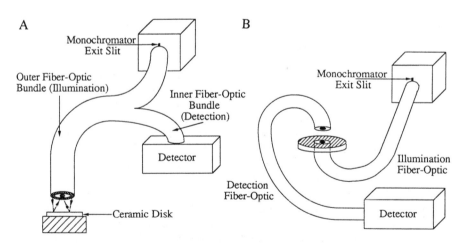

Figure 2A. Reflectance probe design.   Figure 2B. Transillumination probe design.

improve the signal to noise ratio, 64 scans (30 seconds) were co-added. Spectra were acquired in the range of 680-1235 nm. However, only 700-1100 nm were analyzed owing to the poor quality of the data outside this range, resulting from low throughput and poor quantum efficiency.

All data processing and analysis was done in the MATLAB environment (The MathWorks, Inc., Natick, MA). Second derivative transformations used a window (smooth) and gap (derivative) value of 21 data points (roughly 17 nm). Smoothing was accomplished using a running mean (boxcar) smooth, where the value at each data point was replaced by the mean of the values in a seventeen nanometer region surrounding it. After smoothing, second derivative transformations were performed using the method of finite differences. The technique of stage-wise multiple linear regression (MLR) was used to determine which wavelength possessed the best linear correlation of instrumental response to analyte concentration *(20)*. The relationship between known analyte concentration and wavelength of absorbance was displayed as a correlation plot where a magnitude of 1 was indicative of perfect correlation and a value of 0 corresponded to no correlation. Also included on these plots were a relative sensitivity factor, which was essentially the inverse slope of the resulting calibration line. A small sensitivity corresponded to a calibration line with a large slope (i.e., a large absorbance change for a given change in concentration).

Computer Simulations. Computer simulations were performed to aid in the interpretation of experimental results. Moisture gain was simulated by adding successive amounts of a pure water spectrum to a dry cookie spectrum. Changes in oil concentration were simulated by adding different amounts of a spectrum of pure oil to the spectrum of a low-oil cookie. Additive offsets were chosen with the aid of a random number generator. Multiplicative effects were modeled as random changes in the pathlength to approximate experimental observations. The results of the simulation are shown concurrently with the experimental data for comparative purposes.

Results/Discussion

Pure Component Spectra. The four principle ingredients of baked goods are flour, oil, sugar, and water. The SW-NIR transmission spectra of these components are shown in Figure 3. The spectrum of flour exhibits broad, strong absorption bands at 1000 nm and 920 nm. Since the processed commercial flour used in these studies is composed of 75% starch, 10% protein, 1% fat, and 10-15% water, broad bands are expected.

The spectrum of water exhibits a strong band near 975 nm, arising from the $2v_1 + v_3$ combination of stretching motions *(21, 22)*. Although the maximum absorption for water occurs near 975 nm, the second derivative transformation of this spectrum results in a band near 965 nm. Also present are other combination bands near 840 nm $(2v_1 + v_2 + v_3)$ and 750 nm $(3v_1 + v_3)$ *(21, 22)*. These bands are broad due to inhomogeneities in the strength and number of the hydrogen bonds. The bands are also very temperature dependent, because the activation energy for breaking water hydrogen bonds is comparable to kT. The absorption of oil arises from the stretching and bending motions of CH functional groups on the aliphatic chains. The third overtone bands of oil methyl and methylene absorptions occur at approximately 890 nm and 930 nm, respectively. The oil spectrum also exhibits broad bands near 1040 nm arising from the combination of two quanta of stretching and one of bending on the methyl and methylene groups, while the broad bands near 830 nm result from the combination of three quanta of stretching plus one of bending *(23)*. In addition, the fourth overtone absorptions for oil methyl/methylene groups are present at 760 nm . The spectrum of crystalline sucrose exhibits a sharp peak at 980 nm and a smaller and broader band at 920 nm. The sharp absorption bands in the spectrum of sugar constitute an unique signature for crystalline sugar *(24)*. Conversion of crystalline sugar to amorphous or dissolved sugar results in a loss of the sharp band with replacement by a broader feature *(25)*.

A cookie spectrum (Figure 4A) most closely resembles that of flour (Figure 4B), which is its primary constituent. Spectral contributions of the other constituents of baked goods are generally not apparent to the eye due to the predominance of flour absorption. However, owing to the excellent signal to noise, their presence can be reliably detected by multivariate calibration techniques. As will be shown later, the sharp sucrose band at 980 is often noticeable as a shoulder on the 1000 nm flour peak (Figure 27A). The emergence of water is indicated by a filling in of the valley between the 920 nm and 1000 nm flour bands as demonstrated by the spectra of a low and high moisture content cookie in Figure 5A, as obtained by transillumination. The presence of oil is much less obvious, but it does emerge in the 930 nm region at high concentrations as shown by the spectra obtained by transillumination in Figure 5B.

Moisture Determination. Moisture studies were done on commercially prepared vanilla wafer-type cookies which were fairly homogeneous in analyte distribution and uniform in size and shape (Figure 6). The samples were stored in a desiccator for several weeks prior to analysis, therefore, the actual moisture level was several percent less than the typical levels of off-the-shelf product. Moisture levels were varied by suspending the cookies for differing lengths of time in a humid atmosphere produced in a heated vessel containing a small amount of liquid water. Changes in moisture levels of several percent were produced in a few minutes by this means. The samples were allowed to equilibrate to room temperature and then their weight was recorded immediately before and after spectral acquisition. The percent moisture content was calculated from the difference in the average weight of the cookie at the time of spectral acquisition and after oven drying at 130 °C for one hour. Alternate methods of inducing moisture included the use of a constant humidity chamber. Various levels of humidity can be easily achieved with different saturated salt solutions. The use of constant humidity chambers yielded comparable qualitative and quantitative results (slightly higher

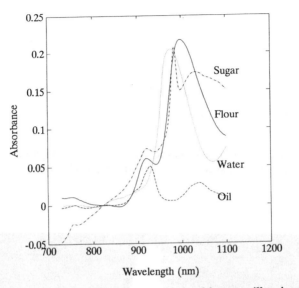

Figure 3. Pure component spectra obtained by transillumination.

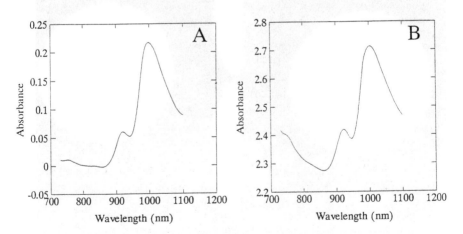

Figure 4A. Transillumination spectrum of flour.

Figure 4B. Transillumination spectrum of cookie.

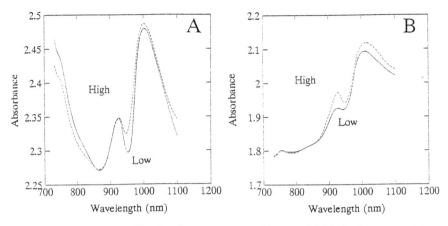

Figure 5. (A) Transillumination spectrum of low and high moisture cookie (B) Transillumination spectrum of low and high oil cookie.

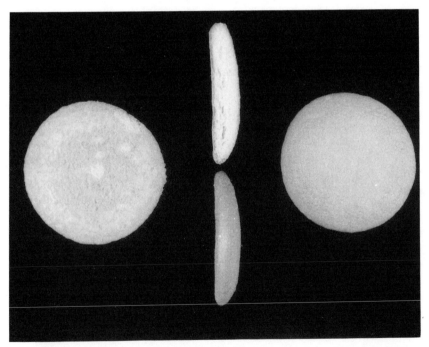

Figure 6. Commercially obtained vanilla wafer type cookies used for humidity studies.

correlation coefficients and lower SEPs) for moisture determination for both reflectance and transmission sampling modes (data not shown).

Transmission. Three cookies with initial moisture contents of 1-1.2% were simultaneously and progressively humidified to approximately 14.5% moisture, resulting in twenty-seven spectra, as shown in Figure 7A. In Figure 8A we exhibit the same data after second derivative transformation. As is well known, this operation removes baseline offsets and slopes and yields higher resolution *(5)*. For comparison, we display in Figures 7B and 8B computer generated spectra produced by successive addition of increasing amounts of a pure water spectrum to the spectrum of a cookie at 1% moisture. In addition, baseline offsets and multiplicative effects were simulated. The parameters of the computer generated spectra were chosen to effect similar baseline offsets and comparable absorbance changes in the simulated data.

The baseline offsets, both additive and multiplicative, arise from inhomogeneity in cookie structure. Variable pathlength effects give rise to multiplicative errors. The general trend of the baseline is downward (more light reaching the detector) with increasing moisture content. However, the variance becomes more random at moisture levels above approximately 7% and obscures this effect. As expected, additive offsets were readily removed by second derivative transformation. Unfortunately, multiplicative effects were unaffected. Instead, these were decreased by normalization of the area of the absolute second derivative spectra. This seemingly arcane procedure can be justified as follows: First, assume that the pathlength term for all wavelengths is the same. Second, assume that to a sufficient approximation the cookie is essentially a two component system consisting of flour and water, and that the flour contribution to the spectrum dominates over the moisture range of interest. Under these circumstances, the ratio of intensity of the water band to that of the flour band can be linearly correlated to the water content independent of the pathlength variation. In practice, we normalized the spectra to unit area rather than taking ratios of intensities because the former procedure yielded a better signal to noise ratio. Thus, the introduction of random noise into the normalization calculation is reduced proportional to the square root of the number of summed channels. Also, the large smoothing functions used in these experiments can cause considerable distortion of peak height and width, while the area of a peak remains relatively constant *(26)*. One drawback to area determination with second derivative spectral data is the precision problem caused by the sign differences of side lobes and main lobe. To circumvent this problem, the area under the second-derivative curve was approximated by the sum of the absolute value of each absorbance.

The second derivative area normalized NIR transmission spectra for the experiment and simulation are shown in Figure 9. To further improve the spectra for interpretability, we display in Figure 10 the second derivative area normalized difference spectra. These result from subtracting the original dry cookie spectrum from all other spectra. This process brings out the otherwise obscure changes arising from water variation. We caution that spectral

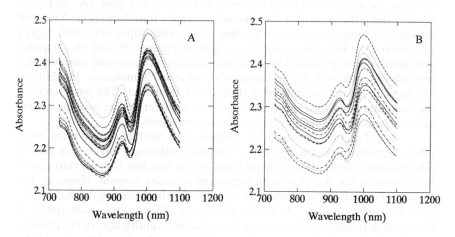

Figure 7. Spectra of humidified cookies obtained in transmission mode (A) Experiment (B) Simulation.

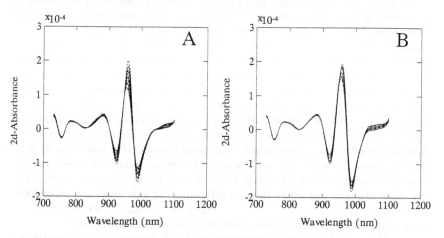

Figure 8. Second derivative spectra of humidified cookies obtained in transmission mode (A) Experiment (B) Simulation.

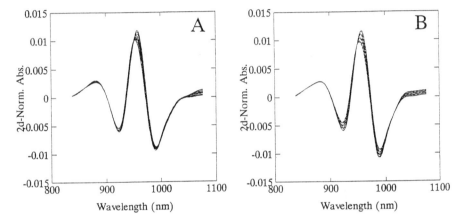

Figure 9. Second derivative area normalized spectra of humidified cookies obtained in transmission mode (A) Experiment (B) Simulation.

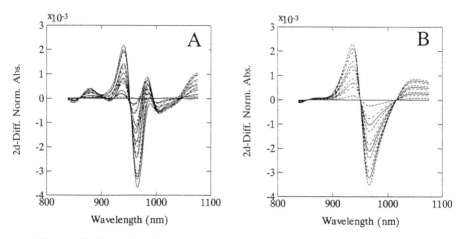

Figure 10. Second derivative area normalized difference spectra of humidified cookies obtained in transmission mode (A) Experiment (B) Simulation.

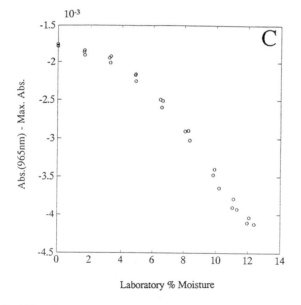

Figure 10C. (Absorbance at 965 nm - magnitude of maximum absorbance) vs. precent moisture for second derivative area normalized humidity data.

subtraction leads to difference spectra which may have several interpretations in terms of the absolute spectra, especially if multiplicative effects are not fully corrected for. In our case, we can confidently state that the difference spectra reveal that the water in the cookies is qualitatively similar to bulk water, but that (a) the spectra are sharper in the cookie and appear to contain more than one absorbing specie near 960 nm and (b) the spectra exhibit a progressive shift of the wavelength of maximum absorbance change to the red with added water. At low concentrations of water, the absorption maximum (which is negative as a result of the second derivative operation) for water occurs near 960 nm in the cookie spectra, while at higher water concentrations the maximum approaches 965 nm, that of pure water. The effect of this change is more clearly shown by displaying the difference between the absorbance at 965 nm and the maximum absorbance (Figure 10C). The simulated spectra show no wavelength shifting of the absorbance band at 965 nm. These observations indicate the presence of more than one form of water-substrate binding, which has been observed by various other techniques *(27, 28, 29, 30)*.

Since the second derivative area normalized difference spectra showed a good qualitative correspondence with added water, MLR was used to assess whether the correlation between selected spectral intensities and known water concentration was sufficiently linear and precise. Accordingly, the correlation of water content as a function of wavelength for the second-derivative area normalized NIR spectra are shown in Figure 11. While there is good qualitative correspondence between data and simulation, subtle variations between the two indicate that adsorption of water by a cookie involves more than simple addition of bulk water, as noted above. Nevertheless, there is good correlation at the wavelengths of known water absorption, namely near 960 nm. By the technique of leave-one-out cross-validation *(31)* it was concluded that the best wavelength to use in this analysis was 961 nm. With this wavelength, a linear correlation coefficient of 0.9980 and an SEP of 0.29% was obtained. The latter corresponds to a coefficient of variation of 3.8% (see Figure 12). While, for these studies, area normalization hardly improved the quantitative results, in the case of the oil determination there was marked improvement, as will be shown. We still prefer area normalization for the water data because the spectra are more interpretable.

Reflectance. Three cookies with initial moisture contents of 1-1.1% were simultaneously and progressively humidified to approximately 14% moisture, resulting in thirty reflectance spectra. The original and second-derivative reflectance spectra are shown in Figures 13A and 14A, respectively. For comparison, we display in Figures 13B and 14B the computer generated spectra. In addition, baseline offsets and multiplicative effects were simulated. Again, the general trend of the baseline in the original spectra is downward with increasing moisture content, becoming more random at moisture levels above approximately 7%.

The second derivative area normalized NIR reflectance spectra and the resulting difference spectra are shown in Figures 15A and 16A. The area

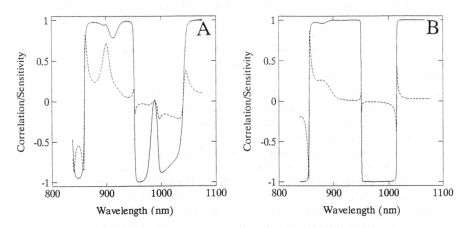

Figure 11. Correlation (-) and Sensitivity (--) vs. Wavelength for humidified cookies in transmission mode (A) Experiment (B) Simulation.

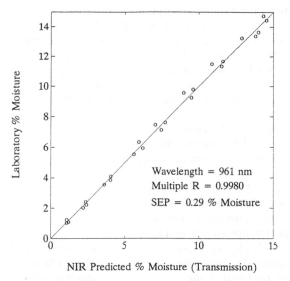

Wavelength = 961 nm
Multiple R = 0.9980
SEP = 0.29 % Moisture

NIR Predicted % Moisture (Transmission)

Figure 12. Actual vs. predicted % moisture using one wavelength linear additive model for humidified cookies in transmission mode.

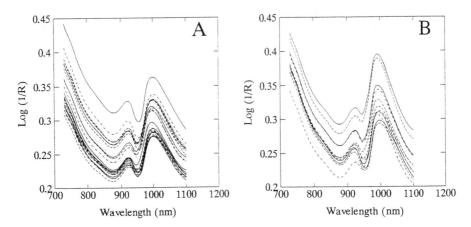

Figure 13. Spectra of humidified cookies obtained in reflectance mode (A) Experiment (B) Simulation.

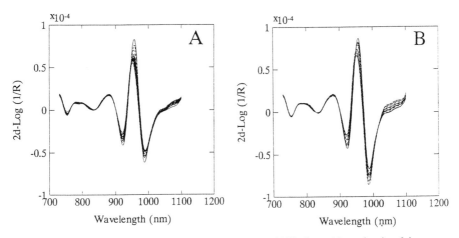

Figure 14. Second derivative spectra of humidified cookies obtained in reflectance mode (A) Experiment (B) Simulation.

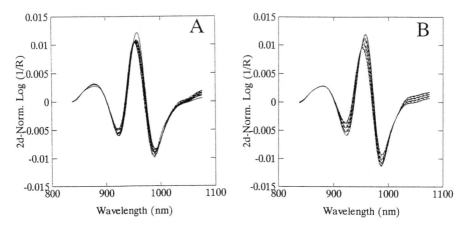

Figure 15. Second derivative area normalized spectra of humidified cookies obtained in reflectance mode (A) Experiment (B) Simulation.

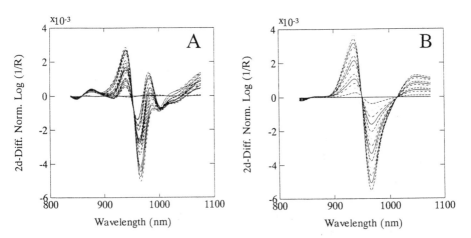

Figure 16. Second derivative area normalized difference spectra of humidified cookies obtained in reflectance mode (A) Experiment (B) Simulation.

normalized spectra and their difference spectra are shown for the simulation in Figures 15B and 16B. The results are nearly identical to the transmission study.

The linear model of MLR was again employed to determine the degree of precision with which known water concentration could be correlated to spectral absorbance change. The resulting correlation of moisture content as a function of wavelength for the second-derivative area normalized NIR spectra are shown in Figure 17. These are to be compared with the multiple R plots of the transmission data in Figure 11. As with the transmission data, there is good correlation at wavelengths near 960 nm, which corresponds to the absorption of water. The wavelength which best correlated spectral intensity with moisture content was 962 nm for the second derivative area normalized data. The resulting SEP for moisture determination using NIR reflectance was 0.31%, which translates to a relative error of 3.7%. A plot of percent moisture determined by oven drying and gravimetry vs. NIR predicted percent moisture using a single wavelength is shown in Figure 18. As in the case of transmission, area normalization did not significantly improve the quantitative results, but it is still preferred for reasons previously mentioned.

Oil Determination. For these experiments, we found that addition of oil to commercial cookies significantly altered their scattering properties and appearance. Accordingly, cookies of different oil content were prepared from batches of dough using five oil levels from 12.5% to 62.5% (weight percent of oil relative to weight of flour). Five batches of cookies were prepared on five different days using two different ovens. The samples ranged in thickness from 3-10 millimeters and possessed various degrees of browning and surface reflection characteristics, as well as numerous irregular features such as holes (Figure 19). The entire batch of samples were allowed to equilibrate a final night at room temperature and humidity before analysis. Conditions of temperature and room humidity were not controlled over the course of the experiment. Reflectance measurements were performed several days after the transmission data was acquired. These conditions represent a rather severe test of the method to perform over a wide variety of formulations and physical characteristics. Such conditions are not likely to be found in a typical processing situation. Thus, significantly better results can be expected on a commercial line.

Transmission. Ten representative spectra, from over one hundred acquired spectra, produced by transillumination are shown in Figure 20 with the simulated data. The thickest samples are distinguished by the largest baseline offsets and absorbance peaks, while the thinnest samples have lower offsets and contrasts. The second-derivative spectra shown in Figure 21 have largely corrected the baseline offset. While similar absorbance excursions are noted between the experimental and simulated data, there was a great deal of variance between the simulated and real absorbance intensities at the wavelength where oil exhibits its most well resolved peak (930 nm). As a

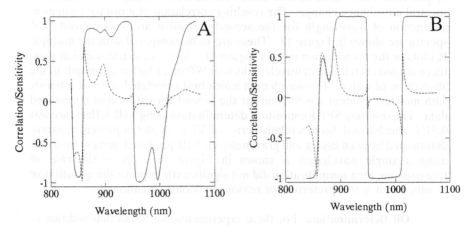

Figure 17. Correlation (-) and Sensitivity (--) vs. Wavelength for humidified cookies in reflectance mode (A) Experiment (B) Simulation.

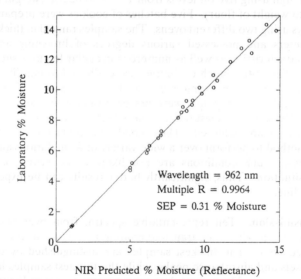

Wavelength = 962 nm
Multiple R = 0.9964
SEP = 0.31 % Moisture

NIR Predicted % Moisture (Reflectance)

Figure 18. Actual vs. predicted % moisture using one wavelength linear additive model for humidified cookies in reflectance mode.

Figure 19. Home made wheat flour cookies used for oil studies.

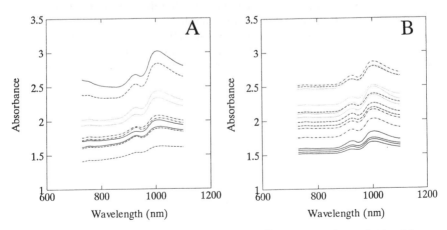

Figure 20. Spectra of cookies with varying oil concentrations obtained in transmission mode (A) Experiment (B) Simulation.

result, very poor correlations were obtained. However, when area normalization was employed to correct for multiplicative scattering effects (i.e., pathlength changes), the simulated and experimental data become much more similar, as shown in Figure 22. In order to better observe the spectral variance resulting from different oil concentrations, the second derivative area normalized difference spectra are shown in Figure 23.

The correlation of oil content with wavelength of absorbance for the second-derivative area normalized spectra is shown in Figure 24. The regions of high correlation include the methylene region near 930 nm, the methyl region near 890 nm, and the combination region near 1000 nm. Using a single wavelength at 927 nm in the second-derivative area normalized data, oil concentration was predicted with an absolute error of 2.5%. This corresponds to a relative error of 6.7% (Figure 25).

Reflectance. Diffuse reflectance spectra were acquired several days after the transmission data. The original log inverse of diffuse reflection spectra are shown in Figure 26. The spectra show a shoulder at 980 nm on the broad 1000 nm absorption band due to flour. Spectra successively acquired in transmission and reflectance mode on the same cookie are shown in Figure 27. The feature at 980 nm is not noticeable in the transmission spectra of cookies. Upon dissection of several cookies, it was noted that a hard layer from one to three millimeters in thickness had formed on the surface. Presumably, this resulted from temperature and humidity fluctuations which caused migration of free water containing dissolved sucrose to the surface of the sample. Upon reaching the surface, the water evaporated and left behind a concentrated layer of sucrose, which then crystallized, yielding the sharp feature at 980 nm *(24)*. Reanalysis of several samples by transmission spectroscopy never revealed the presence of a prominent crystalline sucrose band. The differences between the reflectance and the transmission spectra can be therefore attributed to the bias of reflectance for the top surface of the sample.

The second-derivative spectra for the reflectance experiment and the simulation are shown in Figure 28. As with the transmission data, useful correlation was not possible in the second-derivative. The second-derivative area normalized difference spectra are shown in Figure 29. The simulation results are much more regular than the experimental data, which is again, likely due to varying amounts of crystalline sucrose in the samples. However, the intensity in the 900-950 nm range does exhibit a regular variation with oil content. The correlation of wavelength of absorption with known oil concentration are shown in Figure 30. As expected, it occurs at the known absorption regions for oil. Using a single wavelength of 937 nm, oil was predicted with an absolute error of 2.8%, which corresponds to a relative error of 7.5% (Figure 31).

Conclusions

NIR spectroscopy has proven to be a viable method for real-time oil and water determination of intact cookies. Spectrophotometric analysis of these irregular objects resulted in additive and multiplicative offsets. A second derivative

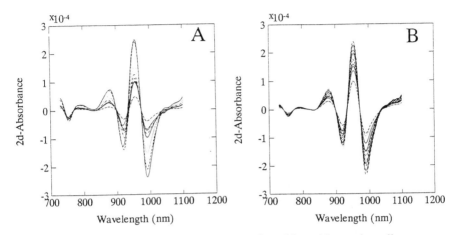

Figure 21. Second derivative spectra of cookies with varying oil concentrations obtained in transmission mode (A) Experiment (B) Simulation.

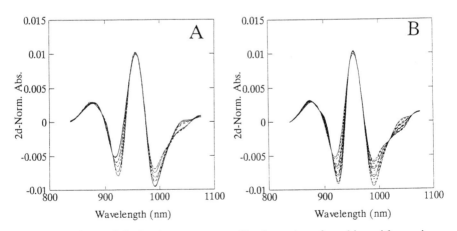

Figure 22. Second derivative area normalized spectra of cookies with varying oil concentrations obtained in transmission mode (A) Experiment (B) Simulation.

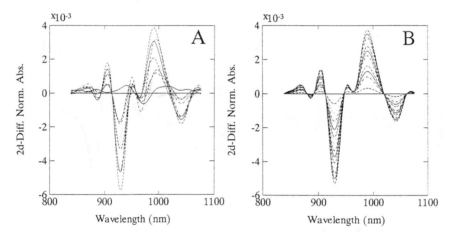

Figure 23. Second derivative area normalized difference spectra of cookies with varying oil concentrations obtained in transmission mode (A) Experiment (B) Simulation.

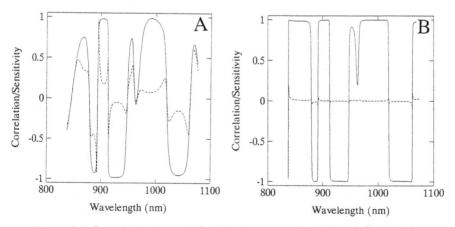

Figure 24. Correlation (-) and Sensitivity (--) vs. Wavelength for cookies with varying oil concentration in transmission mode (A) Experiment (B) Simulation.

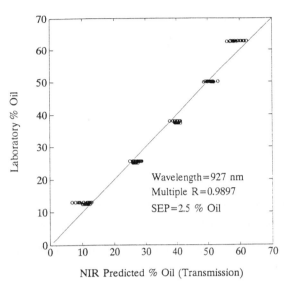

Figure 25. Actual vs. predicted % oil using one wavelength linear additive model for cookies with varying oil concentration in transmission mode.

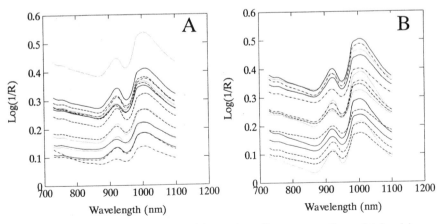

Figure 26. Spectra of cookies with varying oil concentrations obtained in reflectance mode (A) Experiment (B) Simulation.

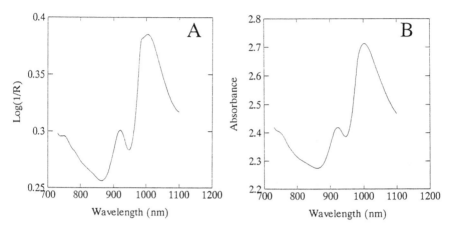

Figure 27. Representative spectrum of cookie (A) Reflectance (B) Transmission.

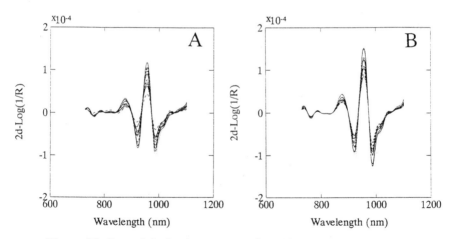

Figure 28. Second derivative spectra of cookies with varying oil concentrations obtained in reflectance mode (A) Experiment (B) Simulation.

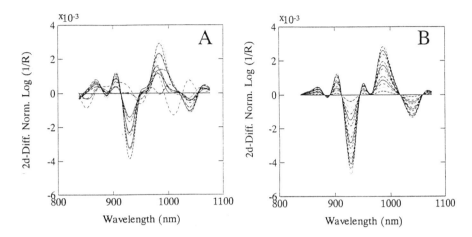

Figure 29. Second derivative area normalized difference spectra of cookies with varying oil concentrations obtained in reflectance mode (A) Experiment (B) Simulation.

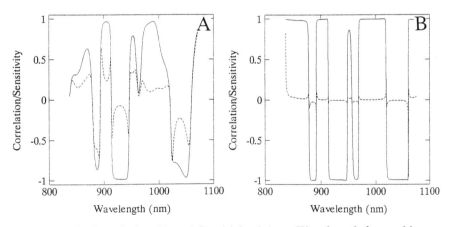

Figure 30. Correlation (-) and Sensitivity (--) vs. Wavelength for cookies with varying oil concentration in reflectance mode (A) Experiment (B) Simulation.

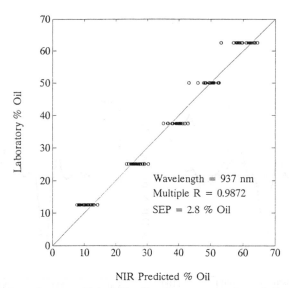

Figure 31. Actual vs. predicted % oil using one wavelength linear additive model for cookies with varying oil concentration in reflectance mode.

transformation readily removes additive offsets, but normalization or wavelength ratioing is necessary to remove multiplicative offsets. Area normalization was found to be preferable over single wavelength normalization. Computer simulations provided a basis on which experimental data could be interpreted.

Using a single wavelength linear additive model near 960 nm, relative errors on the order of 4% moisture were obtained for the analysis of intact cookies for both reflectance and transmission sampling geometries. Further study may reveal that the obvious multi-site binding characteristics of water can be used to effect a more complex model that more precisely predicts water concentration. Also, from the standpoint of basic baking science, NIR appears to have potential for study of free and bound water in baked goods.

Percent oil in whole cookies was predicted with a relative error of 6-7.5% using a single wavelength near 930 nm. The larger relative error for oil determination is primarily due to the extreme differences in sample characteristics. Certainly, oil behaves more independently than water, interacting much less with other baking constituents, as evidenced by comparison of experimental and simulated data. Analysis of smaller subsets within the entire oil determination set reduced relative errors a factor of two when compared to the entire sample set.

For this study, no significant differences were noted for analyses using either reflectance or transmission sampling geometries. However, because of the bias of reflectance for the top surface of a sample, a transmission mode of sampling may be preferred for analysis of thick inhomogeneous objects.

Acknowledgement

This material is based in part upon work supported by the National Science Foundation under Research Experiences for Undergraduates Grant No. CHE-900977.

Literature Cited

1. Lees, R.; *Food Analysis: Analytical and Quality Control Methods for the Food Manufacturer and Buyer;* being the 3rd edition of Laboratory Handbook of Methods of Food Analysis; Leonard Hill Books: London, 1975.
2. *Official Methods of Analysis of the Association of Official Analytical Chemists;* Helrich, K., Ed.; AOAC, Inc.: Arlington, VA, 1990.
3. Pearson, D.; *The Chemical Analysis of Foods;* Seventh Edition; Churchill Livingstone: New York, 1976.
4. Hirschfeld, T.; Stark, E.W.; In *Analysis of Foods and Beverages;* Charalambous, G., Ed.; Academic Press, Inc.: Orlando, Florida, 1984; p.505.
5. *Near-Infrared Technology in the Agricultural and Food Industries;* Williams, P.; Norris, K., Eds.; American Association of Cereal Chemists: St. Paul, MN, 1987.
6. Massie, D.R.; Norris, K.H. *Trans. ASAE.* **1965,** *8,* 598.
7. Baker, D. *Cereal Foods World.* **1985,** *30,* 389.
8. Beebe, K.R.; Kowalski, B.R. *Anal. Chem.* **1987,** *59,* 1007A.
9. Rubenthaler, G.L.; Pomeranz, Y. *Cereal Chem.* **1987,** *67,* 407.
10. Williams, P.; Thompson, B.; Wetzel, G.; Loewen, D. *Cereal Foods World.* **1981,** *26,* 234.
11. Osborne, B.G.; In *Proceedings of the 7th International Symposium on Near-Infrared Reflectance Analysis (NIRA);* Technicon: Tarrytown, New York, 1984.
12. Lysaght, M.J.; Van Zee, J.A.; Callis, J.B. *Rev. Sci. Inst.* **1991,** *62(2),* 507.
13. Mayes, D.M.; Callis, J.B. *Appl. Spectrosc.* **1989,** *43,* 27.
14. Norris, K.H. *NATO Adv. Study Ser. A* **1983,** *46,* 471.
15. Kelly, J.J.; Callis, J.B., unpublished results.
16. Phelan, M.K.; Barlow, C.H.; Kelly, J.J.; Jinguji, T.M.; Callis, J.B. *Anal. Chem.* **1989,** *61,* 1419.
17. Cavinato, A.G.; Mayes, D.M.; Ge, Z.; Callis, J.B. *Anal. Chem.* **1990,** *62,* 1977.
18. Kelly, J.J.; Barlow, C.H.; Jinguji, T.M.; Callis, J.B. *Anal. Chem.* **1989,** *61,* 313.
19. *NBS Standard Reference Materials Catalog 1984-1985;* NBS Special Publication 260; U.S. Government Printing Office: Washington, D.C., 1984.

20. Draper, N.; Smith, H.; *Applied Regression Analysis 2nd Edition;* John Wiley and Sons: New York, 1981.

21. Bayly, J.G.; Kartha, V.B.; Stevens, W. H. *Infrared Physics* **1963**, *3,* 211.

22. Buijs, K.; Choppin, G.R. *J. Chem. Phys.* **1963**, *39(8),* 2035.

23. Kreuzer, J. *Z. Physik* **1941**, *118,* 325.

24. Giangiacomo, R.; Magee, J.B.; Birth, G.S.; Dull, G.G. *J. Food Sci.* **1981**, *46,* 531.

25. Baker, D.; Norris, K.H. *Appl. Spectrosc.* **1985**, *39,* 618.

26. Enke, C.G.; Nieman, T.A. *Anal. Chem.* **1976**, *48,* 705A.

27. Ellis, J.W.; Bath, J. *J. Chem. Phys.* **1938**, *6,* 723.

28. Kuprianoff, J.; *Fundamental Aspects of the Dehydration of Foodstuffs;* Society of Chemical Industry; The Macmillan Co.: New York, 1958; p. 14.

29. Steinberg, M.P.;Leung, H.; In *Water Relations of Foods;* Duckworth, R.B., Ed.; Academic Press: New York, 1974; pp 233-248.

30. Troller, J.A.; Christian, J.H.B.; *Water Activity and Food;* Academic Press: New York, 1978.

31. Wold, S. *Technometrics* **1978**, *20,* 397.

RECEIVED April 3, 1992

Chapter 13

Fiber-Optic Biosensors Based on Total Internal-Reflection Fluorescence

K. R. Rogers[1,3], N. A. Anis[1], J. J. Valdes[2], and M. E. Eldefrawi[1,4]

[1]Department of Pharmacology and Experimental Therapeutics, University of Maryland School of Medicine, Baltimore, MD 21201
[2]U.S. Army Chemical Research, Development, and Engineering Center, Edgewood, MD 21010

A signal transducer based on Total Internal Reflection Fluorescence has been coupled to three biological sensing elements to form receptor, enzyme and antibody biosensors. The nicotinic acetylcholine receptor biosensor is used to detect agonists and competitive antagonists of the receptor with detection limits for the antagonists, which are comparable to those obtained using radioisotope-labeled ligand binding assays. The acetylcholinesterase biosensor is used to detect anticholinesterases. The detection limit for each analyte depends on its potency as an anticholinesterase. Although compounds such as echothiophate and paraoxon yielded detection limits in the ppb range, the biosensor was not particularly sensitive to compounds such as parathion. An immunosensor is also described which uses antiparathion sera as the biological sensing element and detects parathion in the ppb range.

Biosensors are devices which incorporate a biological sensing element with a physical transducer. The optical phenomena of Total Internal

[3]Current address: EMSL, U.S. Environmental Protection Agency, P.O. Box 93478, Las Vegas, NV 89193
[4]Corresponding author

0097–6156/92/0511–0165$06.00/0

Reflectance Fluorescence (TIRF) has been used as the basis for detection in a variety of biosensors (*1, 2, 3*). Since the evanescent wave extends only about 1000 Å into the medium, TIRF is particularly useful for measuring binding events at a solid/liquid interface and can provide the basis for a homogeneous assay (i.e. unbound ligand/antigen need not be physically separated from the immobilized receptor/antibody:ligand/antigen complex). TIRF can also be used to monitor the local environment (e.g. pH, ionic strength, etc.) of a fluorescent probe immobilized to the surface of the waveguide (*4*). In this report we describe receptor, antibody and enzyme applications for fiber-optic biosensors based on TIRF.

Nicotinic Acetylcholine Receptor Biosensor

The nicotinic acetylcholine receptor (nAChR) is the molecular target for a number of neurotoxins, drugs and therapeutics (*5*). Toxins, which bind to this receptor originate from a variety of natural sources. These include α-bungarotoxin (α-BGT), α-*Naja* toxin (α-NT) and α-conotoxin from venoms of the banded krait (*6*), cobra (*7*) and cone snail (*8*), respectively. Other natural toxins include histrionicotoxin produced by the poisonous tree frogs of the Central American rain forest (*9*) and anatoxin produced by the green algae *Anabaena glos-aquae* (*10*). The neurotoxin α-BGT binds to the acetylcholine (ACh) recognition site of the neuromuscular-subtype of nAChR and has been used extensively as a probe for occupancy of the agonist binding site (*11*). The same site binds other competitive antagonists such as α-NT, α-conotoxin and the agonist anatoxin, while allosteric sites on the receptor's channel bind histrionicotoxin. Binding of fluorescein-tagged isothiocyanate (FITC) α-BGT to the nAChR, which is immobilized on the optical sensor, facilitated the detection of a variety of receptor ligands and is the basis for the herein reported nAChR fiber-optic biosensor (Figure 1A).

The nAChR was purified from *Torpedo* electric organ, by far the richest source for this receptor, and noncovalently immobilized to quartz fibers as previously described (*3*). FITC-α-BGT was prepared as described by Rogers et al. (*3*). Nonspecific binding of labeled toxin to the quartz fibers was blocked by treating the nAChR-coated fibers with 0.1 mg/ml bovine serum albumin in phosphate buffered saline (PBS) (154 mM NaCl; 10 mM Na phosphate buffer, pH 7.4), prior to exposure to fluorescent tagged toxin. The fibers were then exposed to 5 nM FITC-α-BGT producing a flourescent signal. In the presence of untagged competing ligands, the fluorescence signal was reduced.

nAChR-coated or noncoated fibers were perfused with FITC-α-BGT at 5 nM in PBS containing bovine serum albumin (BSA) (0.1 mg/ml) (Figure 2). Since the observed fluorescence signal was dependent on the presence of immobilized nAChR, binding of FITC-α-BGT was receptor specific. Binding of α-BGT to the nAChR is a quasi-irreversible event (*11*). Consequently, to insure that equilibrium binding of lower affinity ligands was not altered, initial rates of fluorescence increases were used as indicators of receptor occupancy. Measurements were complete prior to 1% occupancy by the probe of the total available binding sites.

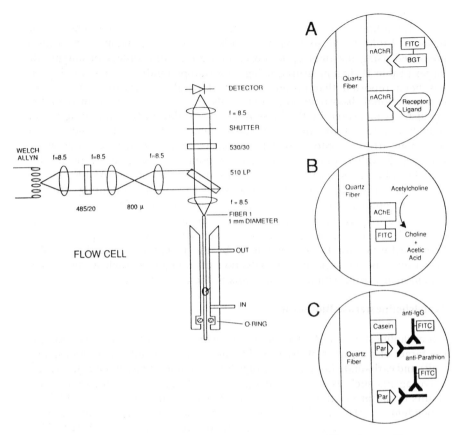

Figure 1. Schematic presentation of the optical system. Fiber insets illustrate biochemical models for (A) nAChR biosensor, (B) AChE biosensor and (C) Parathion (par) immunosensor. Figure adapted from reference (4).

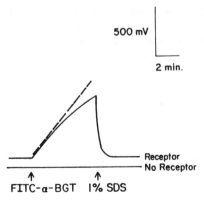

Figure 2. Receptor specific binding of FITC-α-BGT to nAChR-coated or noncoated fibers. Dashed line represents graphically determined initial rates. Receptor and FITC-α-BGT were washed from the fiber with 1% sodium dodecyl sulfate (SDS). Figure was adapted from reference (3).

Inhibition curves were used to calculate K_i values for various cholinergic ligands (Table I), as previously described (15). For agonists, K_i values measured using the fiber-optic biosensor were several orders of magnitude larger (Table I) than measured using radioisotope binding assays (11, 16). It has been reported that proteins adsorbed to glass surfaces show various conformational distortions and constraints (17). The lower affinity of the agonists, as measured by the biosensor, may be due to the limited ability of immobilized receptor to undergo agonist-dependent densitization which results in a high affinity conformation for agonists. For competitive antagonists, however, K_i values were similar for the biosensor to K_d values reported for radioligand binding assays using solubilized preparations of nAChR (11, 18, 19). It is interesting to note that antagonists have identical binding affinities for the receptor in either the resting or desensitized state (20).

The nAChR biosensor could detect agonists of the receptor, such as carbamylcholine and nicotine, with a lower sensitivity than expected from known affinities for these compounds by the receptor. Nevertheless, this biosensor was very sensitive in detection of antagonists such as the snake venom neurotoxins with detection limits in the pmolar range.

Acetylcholinesterase Biosensor

Acetylcholinesterase (AChE), an enzyme which is vital to neural function, is the target of a number of drugs and toxins. These include organophosphate (OP) and carbamate insecticides, drugs for treatment of myasthenia gravis and nerve agents such as soman and sarin (12). Enzymatic hydrolysis of ACh results in the production of acetic acid. Since the quantum yield of FITC is dependent on pH, the FITC-tagged AChE, that was immobilized onto an optical fiber, was used as the basis for an AChE biosensor (Figure 1B). Exposure to various anticholinesterases resulted in a concentration-dependent inhibition of the biosensor response to the substrate ACh.

FITC-AChE was prepared and covalently immobilized onto quartz fibers as previously described (4). The FITC-AChE coated fibers were then placed in the flow cell and perfused with PBS of low buffer capacity (154 mM NaCl; 0.1 mM Na phosphate, pH 7.0). ACh (1 mM) was then added to the flow buffer in the presence or absence of inhibitors. The biosensor assay was performed by interrupting the flow of buffered substrate and measuring the pH-induced fluorescence change over a 2 min period.

Upon interruption of the flow of ACh (1 mM), the pH in the immediate environment of the FITC probe dropped resulting in a decrease in the observed fluorescence (Figure 3). The biosensor signal was enzyme and substrate dependent. Treatment of the biosensor with antiChEs resulted in inhibition of the biosensor response. Inhibition curves for several OP and carbamate antiChEs were used to calculate IC_{50} values (Table II). For the most part, these values correspond well with those determined using a colorimeteric assay of the soluble enzyme (21). Although the biosensor was very sensitive to potent antiChEs such as echothiophate and paraoxon (with detection limits in the ppb range), it was not particularly sensitive to compounds such as parathion and malathion (with detection limits > the ppm range).

Table I Inhibition Constants Measured Using the Fiber-Optic Biosensor and Radioisotope Binding Assays

Ligands	Fiber-Optic Biosensor[a] K_i (M)	Radioisotope Binding Assays K_d (M)
Agonists		
Acetylcholine	1.2×10^{-5}	4.8×10^{-7} [b]
Carbamylcholine	4.5×10^{-3}	6.7×10^{-6} [b]
Nicotine	6.0×10^{-5}	2.5×10^{-5} [c]
Antagonists		
d-Tubocurarine	8.8×10^{-6}	5.0×10^{-6} [d]
α-Bungarotoxin	6.0×10^{-9}	1.0×10^{-10} [b]
α-*Naja* toxin	1.0×10^{-8}	4.0×10^{-9} [e]

[a] IC_{50} values were determined by Log-Probit analysis and K_i values calculated using the relationship $K_i = IC_{50}/(1 + [L]/Kd)$.
[b] Data from reference (*11*)
[c] Data from reference (*16*)
[d] Data from reference (*18*)
[e] Data from reference (*19*)

Table II   Inhibition of Immobilized and Soluble AChE as Assayed by the Fiber-Optic Biosensor and Colorimetric Assays[a]

Compound	Biosensor Assay[b] IC_{50} (M)	Colorimetric Assay[c] IC_{50} (M)
Echothiophate	3.8×10^{-8}	3.5×10^{-8}
Paraoxon	3.7×10^{-7}	4.0×10^{-7}
Bendiocarb	2.2×10^{-6}	6.4×10^{-6}
Methomyl	9.0×10^{-6}	1.5×10^{-5}
Dicrotophos	3.3×10^{-4}	1.1×10^{-4}
Parathion	$> 10^{-3}$	NA
Malathion	$> 10^{-3}$	NA

[a]
[b]   Adapted from reference (4).
[c]   Biosensor assays were performed as described in Methods.
      Soluble AChE was determined by the Method of Ellman et al.   (21).

Parathion Immunosensor

Parathion is widely used as an insecticide in the U.S. until recently and worldwide (13). Although this thiophosphate is not a particularly potent antiChE, its toxic effect is due primarily to metabolic conversion to its oxygen analog paraoxon. Consequently, high affinity recognition of this compound, that is independent of its antiChE activity would be a useful feature for a biosensor.

Several applications of immunosensors based on TIRF have been reported (1, 2). These sensor assays are typically based on either sandwiching an antigen between an antibody immobilized to the sensor surface and fluorescently-tagged antibody used for detection (2), or competition for binding to a fluorescently-labeled antibody between immobilized and soluble antigens (1). We report a parathion biosensor of the latter type (22), which responds to parathion in the 0.1 to 10 µM range (Figure 1C).

Parathion-protein conjugates were prepared as previously described (14). Antisera was produced in adult male white New Zealand rabbits with boosters given every 3 weeks. When antibody titers reached acceptable levels, the rabbits were sacrificed and sera collected, filter sterilized and stored at -80°C. Protein-hapten (50 µg/ml in PBS, pH 7.0) was noncovalently immobilized to the quartz by incubation (4 hr at 4°C). The hapten-coated fibers were then incubated for 8 hr at 4°C in control or antisera (diluted 1:1,000 in PBS) and in the presence or absence of parathion. The fibers were then placed in the instrument. The immobilized antiparathion IgG was then determined by exposure of the fibers to FITC-goat antirabbit IgG (0.5 μg/ml in PBS; Sigma Chemical Co.).

In order to focus on the parathion epitope, casein-parathion conjugate was used as the antigen in the immunosensor assay using antiparathion-BSA from rabbit serum (Figure 1C). The signal arose from binding of FITC-labeled antirabbit IgG to the immobilized parathion-casein/antiparathion IgG complex. The observed signal from the nonspecific binding of FITC-antirabbit IgG to the antigen-coated fiber treated with control rabbit serum was less than 5% of the signal obtained using the antiparathion serum. Addition of parathion to the assay resulted in a concentration-dependent inhibition of the biosensor response (Figure 4). The detection limit for this assay was in the ppb range.

Conclusions

Because the function of complex proteins are typically affected by pH, ionic strength, detergents, and denaturants, biosensors can function only under limited conditions. Nevertheless, receptors, enzymes and antibodies each show distinct binding characteristics, which make them well suited for a variety of biosensor applications. The nAChR biosensor is responsive to a number of drugs and toxicants. Consequently, applications for this type of sensor will most likely be limited to the detection of several groups of compounds which bind to the specific receptor sites or to quantitation of a single agonist or antagonist in media devoid of other related compounds.

Since the AChE biosensor also binds to several classes of inhibitors (i.e. reversible, slowly reversible and irreversible antiChEs), it is also likely to find application in screening samples for compounds such as antiChE insecticides or therapeutics. The ability of the AChE biosensor to perform multiple measurements is a distinct advantage over the nAChR and antibody biosensors, which are primarily limited at present to a single assay per sensor. Since the antiChE-inhibited biosensor can be reactivated by oximes, inactivation of the biosensor by unrelated denaturants such as Hg can be determined.

In contrast to receptors and enzymes, antibodies bind to a single epitope or antigen. As a consequence, immunosensors are well suited for detecting a specific compound in relatively complex media which may contain closely related compounds.

The sensitivity for each of these biosensors depends primarily on the binding affinity of the biological sensing protein for its respective analyte. However, since many of the compounds which are detected by these biosensors bind irreversibly (i.e. α-BGT and OP antiChEs), the detection limits are also dependent on the biosensor exposure time. Consequently, the assay time required by these fiber-optic biosensors can vary from 30 sec to 15 min depending on the analyte. Each of these biosensors show distinct specificity and sensitivity characteristics, which might be exploited for detection of solution analytes under differing circumstances.

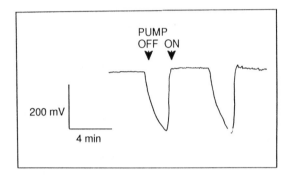

Figure 3. Fluorescence quenching due to AChE activity. In presence of ACH (1 mM), flourescence was quenched when the group was turned off and protons accumulated. The baseline flourescence was quickly reestablished when the pump was turned on and the excess protons were removed by the perfusing substrate solution. Figure was adapted from reference (4).

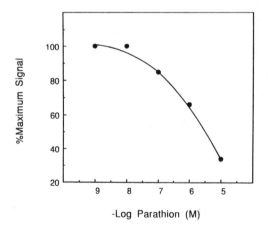

Figure 4. Concentration-dependent inhibition of the parathion-immunosensor response by parathion.

Acknowledgements

This research was supported in part by U.S. Army CRDEC Contract number DAAA-15-89-C-0007 to M.E. and an NRC Research Associateship to K.R.

Literature Cited

1. Kronick, M.N.; Little, W.A. *J. Immunol. Meth.* 1975, *8*, 235-240.
2. Sutherland, R.M.; Dahne, C.; Place, J.F.; Ringrose, A.R. *J. Immunol. Meth.* 1984, *74*, 253-265.
3. Rogers, K.R.; Valdes, J.J.; Eldefrawi, M.E. *Anal. Biochem.* 1989, *182*, 353-359.
4. Rogers, K.R., Cao, C.J.; Valdes, J.J.; Eldefrawi, A.T.; Eldefrawi, M.E. *Fundam. Appl. Toxicol.* 1991, *16*, 810-820.
5. Hucho, F. *Eur. J. Biochem.* 1986, *158*, 211-226.
6. Chang, C.C. In *Snake Venoms*; Editor, Lee, C.Y., Springer-Verlag, Berlin, 1979, pp. 309-376.
7. Albuquerque, E.X.; Eldefrawi, A.T.; Eldefrawi, M.E. In *Snake Venoms*; Editor, Lee, C.Y., Springer-Verlag, Berlin, 1979, pp. 377-402.
8. Olivera, B.M.; Gray, W.R.; Zeikus, R.; McIntosh, J.M.; Varga, J.; Rivier, J.; Santos, V.; Crus, L.J. *J. Science*, 1985, *230*, 1338-1343.
9. Lapa, A.J.; Albuquerque, E.X.; Sarvey, J.M.; Daly, J.; *Witkop Exp. Neurol.* 1975, *47*, 558-580.
10. Carmichael, W.W.; Biggs, D.F.; Gorham, P.R. *Science* 1975, *187*, 542-544.
11. Colquhoun, D.; Rang, H.P. *Molec. Pharmacol.* 1976, 12, 519-535.
12. Taylor, P. In *The Pharmacological Basis of Therapeutics*; Editors, Gilman, A.G.; Goodman, L.G.; Rall, T.W.; Murad, F.; MacMillan Co., New York, N.Y. 1985, pp. 110-129.
13. Zilberman, D.; Schmitz, A.; Casterline, G.; Lichtenberg, E.; Siebert, J. *Science* 1991, *253*, 518-522.
14. Ercegovich, C.D.; Vallejo, R.P.; Gettig, R.R.; Woods, L; Bogus, E.R.; Mumma, R.O. *J. Agric. Food Chem.* 1981, *29*, 559-563.
15. Rogers, K.R.; Menking, D.E.; Thompson, R.G.; Eldefrawi, M.E.; Valdes, J.J. *Biosensors and Bioelectronics* 1991, *6*, 507-516.
16. Eldefrawi, A.T.; Miller, R.; Eldefrawi, M.E. *Biochem. Pharmacol.* 1982, *31*, 1819-1822.
17. Lin, J.N.; Andrade, J.D.; Chang, I.N. *J. Immunol. Meth.* 1989, *125*, 67-77.
18. Moody, T.; Schmidt, J.; Raftery, M.A. *Biochem. Biophys. Res. Commun.* 1973, *53*, 761-772.
19. Johnson, D.A.; Taylor, P. *J. Biol. Chem.* 1982, *257*, 5632-5636.
20. Weiland, G.; Taylor, P. *Molec. Pharmacol.* 1979, *15*, 197-212.
21. Ellman, G.L., Courtney, K.D.; Andres, V. Jr.; Featherstone, R.M. *Biochem. Pharmacol.* 1961, *7*, 88-95.
22. Anis, N.A.; Wright, J.; Rogers, K.R.; Thompson, R.G.; Valdes, J.J.; Eldefrawi, M.E. Analyt. *Anal. Lett.* 1992, *25*, 627-635.

RECEIVED May 27, 1992

Chapter 14

Needs for Biosensors in Space-Biology Research

Sjoerd L. Bonting

SETI Institute, National Aeronautics and Space Administration–Ames Research Center, Moffett Field, CA 94035

Considerable advantages are expected from the use of biosensors in space biology research. They provide (near) real time monitoring of many important bioanalytes. Ideally, it is possible even to achieve continuous monitoring. Their use can save a considerable amount of crew time, which is always in short supply. The sensor output can without much difficulty be made available to the ground by telemetry. Desirable types of bioanalytical sensors for space biology research and monitoring of the quality of recycled water during long-term missions are discussed.

Carrying out traditional bioanalytical procedures during spaceflight is complicated for several reasons. Sample treatment is more difficult in a weightless environment than on earth. Use of toxic reagents can only be permitted under conditions of double containment. Analytical instruments take up space and use power. The usual bioanalytical procedures are rather time-consuming. Crew time, pressurized volume, and power are always scarce resources in space missions.

During the relatively short missions of the Space Shuttle onboard analysis is mostly avoided by performing tests immediately before and after flight, in humans as well as in animals; in the latter case there are also ground control animals. Any samples obtained during flight are preserved and analyzed after return to the ground.

Need for Onboard Analysis in the Future

With the advent of the Space Station Freedom, onboard analysis of a large number of analytes can hardly be avoided. Experiments of months duration will be performed to study the long-term effects of space conditions on living

0097–6156/92/0511–0174$06.00/0

organisms. In such experiments it will be very important to be able to measure various biochemical parameters at frequent intervals during the experiment in order to monitor the course of adaptation of the organism to space conditions. However, logistics flights of the Space Shuttle will be at intervals of at least three months. This means that the analytical results on samples returned to earth would become available with several months delay, which is obviously undesirable. Thus the ability to perform onboard analysis of a variety of biochemical parameters becomes very crucial for space biology research on Space Station Freedom. This will also be true for other long-term missions in the future, like a manned mission to Mars or a permanently manned Lunar base.

During such long-term missions there will also be a need for a closed or nearly closed environmental life support system (CELSS), which provides regeneration of the gas atmosphere and the water used by crew and other living organisms. This will also require the monitoring of a number of chemical parameters. If the monitoring can be carried out continuously, then this provides the possibility of optimizing and controlling the regeneration processes.

Advantages of Biosensors for Onboard Analysis in Spacecraft

The use of chemical sensors offers considerable advantages over conventional analysis onboard spacecraft. In principle, chemical sensors can provide (near-) real time monitoring of many important analytes. In some cases they can even provide continuous monitoring of such analytes. The sensors, and even the ancillary instruments, are small compared to conventional analytical instruments. Their power consumption is low.

Sensor measurements do not require extensive sample treatment before analysis; in many cases a biosensor can simply be inserted in or be placed on the organism. In the case of CELSS processes, sensors can be placed in the flow of air or water. This means that far less crew time is needed than for conventional methods of analysis. The use of sensors thus provides an efficient use of the scarce resources of crew time, pressurized volume, and power.

Space experiments are often monitored, and even controlled, by scientists on earth. This requires that data be transmitted to the ground, preferably near real time. Here sensors have the advantage that their output is usually in digitized form, which lends itself to rapid transmission to the ground. If the sensors are made to transmit their output by telemetry, then this has the additional advantage of allowing free movement of the experimental subject.

For these reasons, NASA has initiated a program devoted to the study and development of chemical sensors for spaceflight (1). Table I shows the scope of this program. Investigators can submit a proposal for a sensor which offers promise for application in spaceflight and may then receive a contract from NASA for its development.

Sensor Requirements for Spaceflight Use

Obviously, sensors to be used for operation during spaceflight have to satisfy a number of requirements, some dictated by the conditions peculiar to the space environment, others to ensure the safety of the mission and of the crew onboard. The requirements are listed in Table II.

"Operating in microgravity" means that the functioning of the sensor does not depend on gravity. "Accurate" means not only that the analyte concentration is provided within a certain percentage error, but also that the specificity of the sensor is such that no serious interference from other substances present in the sample occurs. "Stable" means that the sensitivity of the sensor does not change appreciably during the period of measuring.

In most cases the sensor produces a signal in the form of an electric current or voltage that requires calibration in terms of analyte concentration. The calibration process should be easy, requiring little crew time. Automatic calibration is, of course, preferable; it can in some cases be achieved by the use of a reference sensor inserted in a standard solution.

"Long lifetime" refers both to shelf-time of the sensor, and to the continued functioning of the sensor during an experiment of long duration. Ideally, a sensor should not have to be replaced during the course of an experiment.

"Biocompatibility" is, of course, a strict requirement for all invasive sensors, those that are to be used inside the human or animal body. This does not only mean that the sensor must be constructed of non-toxic, non-corroding

Table I. Scope of the 'Sensors 2000!' Program of NASA

- Development of physical, chemical, and biological sensors for spaceflight;
- Advanced sensor development;
- Spaceflight hardware development;
- Biotelemetry system development;
- Workshops, conferences, symposia, consortia.

Table II. Sensor Requirements for Spaceflight Use

Operating in microgravity	Minimal size and weight
Accurate	Low power consumption
Stable	Minimal complexity
Easy or automatic calibration	Easy to service
Long lifetime	Materials space-qualified
Biocompatibility (for invasive sensors)	

materials. An as yet unsolved problem with invasive sensors is that eventually a series of tissue reactions occurs which increasingly impairs the function of the sensor.

The requirements for minimal size and weight and low power consumption will be clear from what has been said in the previous sections. Minimal complexity and easy servicing are requirements typical for spaceflight conditions, where crew time is a precious commodity and crew members cannot be expected to be experts in the use and repair of each kind of sensor in addition to all other equipment onboard.

All materials onboard a NASA spacecraft must conform to the requirements set by NASA in terms of structural stability, flammability, explosiveness, and toxicity (2). Extensive testing is required before any equipment is space-qualified.

Sensor Technologies Currently Used in Space Missions

Sensor technologies used in space biology research so far are shown in Table III. The first four missions were unmanned satellites, Biosatellite III of NASA and Cosmos of the Soviet Union. The animals carrying biosensors on these missions were small primates (Macaca and Rhesus). The fifth one was a Shuttle/Spacelab mission carrying an improved version of the Research Animal Holding Facility (RAHF) with rats. The duration of these mission ranged from 5-13 days.

It will be noticed that only biophysical sensors were used in these missions. This is because these sensors have been available for some time,

Table III. Sensor Technologies in Current Use

BIOSATELLITE III	1969	Temperature, pressure, pO_2, pCO_2 in capsule Food and water consumption EEG, EOG, EMG, brain temperature ECG, respiration Venous and arterial pressure catheters
COSMOS 1514	1983	Temperature and activity sensors Carotid pressure and flow cuff
COSMOS 1667	1985	Carotid pressure and flow cuff
COSMOS 20441	1992	EOG, EMG Activity and skin temperature sensors
SPACELAB SLS-1	1991	Implanted sensors for body temperature, venous pressure, and aortic blood flow

abbreviations: EEG - electroencephalogram, EOG - electro-oculogram, EMG - electromyogram, ECG - electrocardiogram while biochemical sensors are still under development. A further point is that implanted sensors were used on the NASA missions, which was possible because of the relatively short mission duration.

Under development by NASA are biosensor systems for the Cosmos 1992 primate flight experiment, the SLS-2 cardiovascular rat experiment, and the SLS-3 rhesus research facility. For the Cosmos 1992 mission vestibular neural response (VNR), EEG, EMG, and EOG measurement systems are being developed, which make use of modular hybrid electronics. The rhesus project will use implanted ECG and deep body temperature transmitters, and hardwired, surface-mounted EMG and EEG measurement systems. The SLS-2 mission will monitor cardiac output by means of pulsed Doppler ultrasonic flow measurement and arterial pressure, both using implanted biotelemetry systems.

A prototype ionic calcium sensor using a coated wire electrode with reference electrode has been developed, which is to be incorporated in a totally implantable biotelemetry system for chronic studies of untethered animals (3).

Types of Biochemical Sensors Needed

Table IV lists a number of biochemical sensors that would be useful for space biology experiments. The list is not exhaustive; eventually nearly every biochemical parameter would be of interest.

The fact that many of these sensors would also be useful for medical diagnostic purposes could assure a sizable market on earth, potentially much larger than that for space research purposes.

Not all biochemical sensors need to be implantable. E.g., for the analysis of urine samples, the sensor could be dipped into the fluid. For blood analysis an implanted or a skin sensor is preferable. Where this is technically not feasible, the alternative would be to draw a blood sample and to place a drop of blood on the sensor. The latter procedure is being followed with currently available blood glucose sensors for diabetics.

Table IV. Analytes For Which Chemical Sensors Are Needed

Ions Na^+, K^+, Ca^{2+}, Mg^{2+}, Cl^-, PO_4^{3-}, HCO_3^-, F^-, pH

Gases O_2, CO_2, H_2, CO, NH_3

Metabolites Glucose, lactic acid, creatinine, cholesterol, etc.

Enzymes   Alkaline phosphatase, SGOT, CPK, LDH, etc.

Hormones   ACTH, ADH, adrenaline, aldosterone, cortisol, growth hormone, etc.

Total Organic Compounds and Microbial Count in recycled water

Biochemical Sensors with Proven or Potential Usefulness

Ions can be determined by means of ion-selective sensors, consisting of an ion-selective membrane and a potentiometric device. Ions that can thus be determined are: Na^+, K^+, Li^+, H^+, Cs^+, NH_4^+, Ca_2^+, Pb_2^+, Cl^-, HCO_3^-, NO_3^-, ClO_4^-. The membranes are made ion-specific by the incorporation of an ion exchanger or an ionophore (4, p. 197-276), or they consist of an ion-selective glass coating on a glass electrode (5, p. 135-140). The calcium sensor, mentioned in the previous section, is an example of the latter type. Another possibility is to use an ion-selective field effect transistor (ISFET), although there are still problems with long term baseline drift and poor adhesion of the ion-sensitive membrane (6, p. 171-174).

Gases can be determined in a variety of ways. For respiratory gases a space qualified mass spectrometric technique is available, which is probably the most universal method for gas analysis. For blood gases (O_2 and CO_2) a Clark polaro-graphic electrode is available for insertion into a blood vessel, and also an adapted version for placement on the skin for transcutaneous measurements (5, p. 356-364). For hydrogen and ammonia a metal oxide semiconductor field-effect transistor (MOS-FET) offers possibilities (4, p. 528; 5, p. 522-528; 6, p. 174-176). A microbial biosensor, consisting of an immobilized microorganism and an electrochemical device, has been developed for carbon dioxide and methane (6, p. 166-167); the drawback is its limited lifetime. Volatile organic compounds, like n-pentane, acetone, isopropanol, chloroform, and toluene, can be determined by means of an etched surface acoustic wave device (SAW), which detects mass changes due to gas adsorption as a proportional shift in resonance frequency of a piezo-electric crystal (7). Finally, there is a potential for the use of powder-based semiconductor gas sensors for a variety of gases (O_2, CO, H_2, NO_2, hydrocarbons), adsorption of which causes a change in resistance (4, p. 479-516).

For small molecular metabolites the use of immobilized enzyme sensors is a possibility, although antibodies specific for such compounds have also been developed. A crucial point in the development of enzyme sensors is the immobilization of the enzyme without appreciable loss of activity (5, p. 85-99). The transducer can be an electrochemical device. An example is a glucose sensor, with immobilized glucose oxidase which oxidizes glucose to gluconic acid and H_2O_2. The resulting local pH decrease can be measured with a pH glass electrode, the H_2O_2 formed can be determined amperometrically (4, 285-294), or potentiometrically with a co-immobilized glucose oxidase/catalase sensor (6, p. 198). Other metabolites that can be determined by enzyme sensors are: lactate, creatine, ATP, urea, amino acids, cholesterol (6, p. 19-46).

For enzymes and hormones immobilized antibodies are the most suitable sensor material, although isolated hormon receptors are used for some hormones. The transducer can be a) an electrochemical device, b) an evanescent wave optic fiber system, c) a surface plasmon resonance (SPR) device, or d) a surface acoustic wave (SAW) device. An example of a) is the use of a catalase-labeled antigen. When the catalase-labeled antigen binds to the antibody, which is immobilized on a platinum electrode, the reduction of added H_2O_2 is measured amperometrically. Native antigen (the analyte) displaces labeled antigen and leads to a decrease in the amperometric reading proportionally to its concentration (6, p. 299-301). In b) use is made of total internal reflection in an optic fiber with light penetrating to a depth of about 0.5 wavelength into the surrounding fluid (evanescent wave). Coating the fiber with an antibody layer, to which fluorescent-labeled antigen is bound, will introduce fluorescence light into the fiber which can be measured by a photometric device. The native antigen analyte will displace part of the labeled antigen, thus reducing the fluorescence proportionally (5, p. 655-661; 8). In c) the intensity of light totally reflected in a glass prism coated with a thin metal layer (60 nm gold or silver), on which an antibody layer is deposited, is measured as a function of the angle of incidence. At a certain angle a sharp minimum in reflection occurs. When antigen binds to the deposited antibody, the reflection minimum shifts to a lower angle, proportionally to the amount of antigen bound (5, p. 661-663). In d) binding of antigen is measured as a shift in the resonance frequency of the piezo-electric crystal.

Sensors for Monitoring Water Recycling

The purity of recycled water is of vital importance for the crew during long term space missions, as on Space Station Freedom, a Lunar base or a Mars mission. Requirements for potable and hygiene water have been set by NASA (Table V). Various parameters will need to be monitored, such as pH, conductivity, color, turbidity, total organic carbon (TOC), microbial count, presence of organic toxics (dichloromethane, butanol-1, ethanol, m-xylene, methyl ethyl ketone, acetone, propyl acetate, halon 1301, methyl isobutyl ketone, propanol-2, butanal, cyclo-hexane, toluene, cyclohexanol, methane, methyl acetylene, 1 11-trichloroethane, methanol) and inorganic toxics like ammonia and heavy metals. On-line monitoring would be the preferred method, although for less critical parameters periodic monitoring (once per 1-4 weeks) would be acceptable. A schedule according to which the monitoring could be conducted is shown in Table VI.

New in this list of parameters are TOC and microbial count. Since TOC may comprise such diverse compounds as aldehydes, alcohols, hydrocarbons, plasticizers (released by plastics), soap and detergents, a non-discriminating sensor will be needed. This might be accomplished with an evanescent wave fiber optic probe where the adsorption of organic molecules causes a change in refractive index at the fiber surface and thus a change in the intensity of light passing through the fiber. Adsorption of a variety of organic compounds can be facilitated by making the fiber surface hydrophobic through silanization.

For microbial counting some form of concentrating will be necessary in order to provide sufficient sensitivity of the microbial detection. For this purpose a bypass with a microbial filter at the entrance as well as at the exit of the storage tank could be used. At periodic intervals the filter could be removed and placed in the assay vessel. The presence of microbes might be detected with the light-addressable potentiometric device (LAPS), which is based on the principle that urease in the presence of DNA will release hydrogen ions that are measured potentiometrically. If microbial growth is detected, it would probably be necessary to diagnose the species present. For this purpose there is available a testkit with differential media to which fluorogenic or colorigenic compounds have been added. The filter on which the microorganisms have been collected is cultured, and the results can be available in 18-24 hours. The phage method and the gene probe technique are faster and more sensitive, but are more laborious and less suited for the space environment.

Inorganic ions could be determined with ion-selective electrodes (9), preferable in multiple probe arrangement (10). For metals which form amalgams the potentiometric stripping method could also be used (11).

Table V. Maximum Contaminant Levels (MCL)
for Potable Water and Hygiene Water

Parameter	Potable	Hygiene	Parameter	Potab	Hygie
Physical			Bactericides, mg/l		
Total Solids, susp./diss.,mg/l	100	500	Residual Iodine, min.	0.5	0.5
ColorTrue,Pt/Co	15	15	Ibid.,max.	4	6
Taste and odor,TTN/TON	3	<3			
pH	6 - 8	5 - 8	Inorganics, mg/l		
Particulates,max.size,*m	40	40	Cations	30	
Turbidity,NTU	1	1	Anions	3	
Dissolved gas (free at 35°F)	none	none	CO2	15	
Free gas (at STP)	none	none			
			Microbial		
Inorganic constituents, mg/l			Bacteria,total count, mg/l	10	10
Ammonia	0.5	0.5	Anaerobes	10	10
Arsenic	0.01	0.01	Aerobes	10	10
Barium	1.0	1.0	Gram Positive	10	10
Cadmium	0.01	0.01	Gram Negative	10	10
Calcium	30	100	E. coli	10	10
Chloride	250	250	Enteric	10	10
Chromium	0.05	0.05	Viruses, PFU/l	10	10
Copper	1.0	1.0	Yeasts and molds, CFU/l	10	10
Iodine, total	15	15			
Iron	0.3	0.3	Organics, µg/l		
Lead	0.05	0.05	Total acids	500	TBD
Manganese	0.05	0.05	Cyanide	200	TBD
Magnesium	50	50	Halogenated hydrocarbons	10	TBD
Mercury	0.002	0.002	Phenols	1	TBD
Nickel	0.05	0.05	Total organic carbon(TOC)	500	TBD
Nitrate,NO3-N	10	10	TOC,less non-toxics	100	TBD
Potassium	340	340	Specific Toxics	TBD	TBD
Selenium	0.01	0.01	Total Alcohols	500	TBD
Silver	0.05	0.05			
Sulfate	250	250			
Sulfide	0.05	0.05	Radioactive constituents to conform to		
Zinc	5	5	Fed. Register, vol.51,no.6, 1986, App.B		
Flouride	-	1.0	Table 2,col.2		

Source: ref.12.

Additional maximal levels in mg/l in potable water: Sodium 150, Carbonate 35 (ref.13).

Table VI. Water Quality Monitoring Schedule for Potable Water and Hygiene Water

Parameter	On-Line	Batch	Periodic
Conductivity	x	x	-
pH	x	x	-
Turbidity	-	x	-
Color	-	x	-
Temperature	x	-	-
Ammonia	x	x	-
Iodine	x	x	-
Selected specific ions	-	x	-
Other inorganic constituents	-	o	o
Totalorganic carbon (TOC)	x	x	o
Selected organic constituents	-	x	o
Total bacteria	-	x	
Yeasts and molds	-	-	x
Bacterial identity	-	-	x
Viruses	-	-	-
Radionuclides	-	x	o
Dissolved gas	-	x	-
Free gas	x	x	-

x monitoring required
o monitoring requirements and schedule dependent on water use and recycling process
- monitoring not required

Source: ref. 12.

References and Notes

1. This is the 'Sensors 2000!' program, managed by John W. Hines, NASA-Ames Research Center, Code 213-2, Moffett Field, CA 94035 (tel.:415-604-5538; fax: 415-604-4984). A Newsletter is available from this office.
2. Safety Policy and Requirements for Payloads using the Space Transportation System, NASA, Washington, D.C., 1980.
3. Newsletter of the NASA-Ames Sensors 2000! Program, vol. 1, no. 1, p. 5,September 1990.
4. M.J. Madou and S.R. Morrison, Chemical Sensing with Solid State Devices,Academic Press, Boston, 1989, 556 pp.
5. A.P.F. Turner, I. Karube, G.S. Wilson, eds., Biosensors, Fundamentals and Applications, Oxford University Press, Oxford, 1989, 770 pp.
6. A.E.G. Cass, ed., Biosensors, A Practical Approach, IRL/OUP, Oxford, 1990,271 pp.

7. M. Thompson, U. of Toronto, Canada, personal communication, July 1991. For the SAW technique, see ref. 5, p. 551-559).
8. R.B. Thompson and F.S. Ligler, Chemistry and Technology of Evanescent Wave Biosensors, in: Biosensors with Fiberoptics, Wise and Wingard, eds., Humana Press, 1991 , p. 111-138.
9. A Guide to Ion-Selective Electrodes, D.C. Hixon, Nature 33S, 379-380,1988.
10. Dr. Jeffrey Daniels, Lawrence Livermore National Laboratory, Livermore,CA, tel.: 510-422-0910, fax: 510-423-6785.
11. P. Hansen, Heavy Metal Determination Using Potentiometric StrippingAnalysis, American Laboratory, April 1991, p. 52-58.
12. Man-Systems Integration Standards, NASA-STD-3000, vol. 1, revision A, NASA, Washington, D.C., October 1989, pp. 7/5-7/24.
13. Cal C. Herrmann, Physical/Chemical Closed-Loop Life Support Systems: Water Quality, Standards, Specifications, and Methods, Project Report, September 1989, NASA Project "Pathfinder", Contract NAS2-12346, NASA-Ames Research Center, Moffett Field, CA 94035, 1989, 59 pp.

RECEIVED May 27, 1992

Chapter 15

Mountaineering in Research and Development

Bego Gerber

Technology Management Group, Business Development International,
1505 Masonic Avenue, San Francisco, CA 94117

This chapter briefly discusses the current business climate, then introduces the mountaineering paradigm. This concept is applied to the relationship between R&D and break-even. The experiment budget concept is introduced. Mountaineering is applied to R&D, and the practical consequences for experimental design are shown to improve the company's bottom line.

The Business Climate

Japan, South Korea, Taiwan, and Hong Kong have become major forces in the world marketplace. In addition, many Eastern European countries are soon to join them, not to mention Central and South American countries, Portugal, and the People's Republic of China. Many of these countries have the advantage of an educated (or educable) and relatively inexpensive workforce.

Countries like the U.S. have not been competing well. Unfortunately, the record shows that many U.S. companies do a less than wonderful job of getting their products into the marketplace profitably. This situation is very clearly so in, for example, the bioscience arena, despite some obvious successes.

We also tend to point at others as the causes of our own difficulties. Typical explanations are "the venture capitalists decimated us", the FDA delayed the approval", and "the recession hurt us".

The following kind of scenario is common in many companies. A device goes from the Development Department to Manufacturing and seems to perform poorly there, or the manufacturing process fails. Development says, "Those folk in Manufacturing always ruin things." Meanwhile, Manufacturing says, "Development always gives it to us before it is ready." Often, both departments are correct. A better outcome was available: if both departments had an integrated approach to product development, the barriers between them would be much smaller, and the process much smoother. Ultimately, it is management's responsibility to institute the mechanisms that ensure that everyone in the company is climbing the same mountain.

An obvious strength of the United States is its science and technology. This strength is evidenced by the size of meetings like the ACS national meetings and the quality of professional publications. We are all very bright people. Intelligent, effective solutions to problems is what we are especially well trained to accomplish.

0097–6156/92/0511–0184$06.00/0

Lots of excellent tools and techniques are available to us to facilitate problem solving. Many of our existing competitors are already using them. In fact, I have just returned from a trip to the People's Republic of China to advise the Chinese government on the development of their new export-oriented Special Economic Zones. I can assure you that they will be taking advantage of these tools without hesitation. In order to improve things in the face of such competition, we must change the way we approach some of the critical issues before us. How can we accomplish that in a manner consistent with our own backgrounds and ways of operating?

The Mountaineering Paradigm

Let's suppose you are an employee of HealthCo, a growing healthcare company. You could be anywhere in the organization, from the Custodian to the Chair of the Board. Imagine how it would be if your products went out the door on time every time, with excellent quality, and made lots of money. This set of circumstances is especially appleaing because HealthCo is well known for its outstanding profit-sharing incentives program.

Mountaineering is a valuable paradigm for accomplishing such an end and is just one example of countless potentially beneficial paradigms.

The dictionary defines a paradigm as an accepted example, model, or pattern. Loosely paraphrasing Thomas Kuhn (*1*), a paradigm is a set of rules that defines problems and their solutions. Kuhn says (p. 37): "... one of the things a scientific community acquires with a paradigm is a criterion for choosing problems that, while the paradigm is taken for granted, can be assumed to have solutions." He also describes (p. 35) "normal science as puzzle-solving" and mostly mopping-up operations". In part, what Kuhn is saying is that we choose problems we think we can solve. Thus the adage: "If all we have is a hammer, we treat everything like a nail."

Statisticians address a related idea when they talk about errors: Errors of the First Kind occur when we claim differences that do not in fact exist (*2*); Errors of the Second Kind occur when we fail to find differences that really do exist (*2*); and Errors of the Third Kind occur when we get the right answer to the wrong problem (*3*). This third kind of error is related to Kuhn's definition of a paradigm.

To return to the concept of mountaineering, imagine it is May 1953, and you are with the New Zealander Edmund Hillary and the Sherpa Tenzing Norgay on one of several teams trying to scale Mount Everest for the very first time in history. Many have already died in prior attempts. The odds against you are enormous, the competition is fierce, the climate is extremely forbidding and getting worse, and your supplies and human resources are quite limited. Nevertheless, you really believe you have a reasonable chance of getting to the top, and of doing so before everyone else. You are full of courage and determination, you're doing something you love, and your eyes glaze over at the promise of fame and fortune, and the prospect of a supreme sense of personal accomplishment.

By the mountaineering paradigm I mean the approach to problem-solving that considers the problem as a mountain to be climbed. It is not a major conceptual step to substitute the successful development and introduction of HealthCo's products into the marketplace for the climbing of Everest. You know that danger lurks at every step and that time is of the essence. It needs good planning, good technique, adequate resources, and a healthy dose of good luck. Thus mountaineering is a relevant paradigm for product development.

The Profit Mountain

Is Mountaineering relevant to HealthCo in any other ways?   One of HealthCo's goals is to make profits by selling diagnostic products that the customer wants to pay for. We will all agree that profit is paramount in our business. Without profit, or at least the promise of profit, we would not have dinners to eat. The reason the employees are at HealthCo is to increase the company's profits as quickly and consistently as possible.

Imagine now a mountain plotted on three axes: sales, expenses, and profit (Figure 1). A company's goal is to reach the top of the profit mountain as quickly as possible. Furthermore, until there are products on sale at a profit, the company can't even begin to climb up the mountain at all, because expenses never go away. The sooner we start selling, the better—with some caveats, of course. Indeed, we are willing to spend a lot of money to reach that point; and, just like Edmund Hillary at Everest, it is hard to justify all the expense unless you make it to the peak in time.

This argument could apply to any department, group, or individual within a company. It is worthwhile to ask ourselves as we meet different people: "What mountain is she or he climbing? What are the factors (graphical axes), the climate, etc.?" This question is very valuable because one of the biggest barriers to success in companies is the inadequate communication between groups and departments. (Remember the Development versus Manufacturing conflict mentioned.)

Break-even Analysis

It is also fruitful to look at HealthCo's profits in terms of an economist's break-even analysis. In the break-even analysis diagram (Figure 2), profit is the difference between the net revenues and the net costs. The net cost is the sum of fixed costs and variable costs. The variable costs are those generated directly through the sale of a unit of product, such as shipping and handling. The fixed costs are those associated with running the business regardless of how many units are sold, such as maintaining sales and shipping departments. Revenue is all income from the sale of the product.

The break-even point is where the net revenue and net cost lines of the graph intersect. Profit is the "quadrant" to the right of the break even point, between the two lines. As you can see from the graph, the lower the fixed cost associated with each unit, the fewer units must be sold to break even and be profitable. The result is increased profitability: these additional profits can then go to bonuses in cash or stock, or to improving facilities, or a variety of other places that can make the company happier, more productive, and even more profitable.

Both the fixed and variable costs are, therefore, the sums of many component costs. (Consequently they both vary as a function of time.) In principle, all the expenses of running the business are included in these costs. Research and development costs are among these, as part of the fixed costs. (Sometimes, misleadingly, R&D costs are included in "overhead", which itself should be

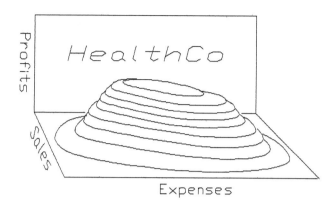

Figure 1. HealthCo's profit mountain.

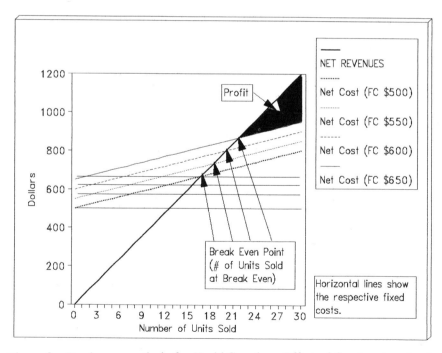

Figure 2. Break-even analysis for HealthCo sales: Effect of fixed cost on break-even point.

considered a fixed cost; sometimes R&D is simply called R&D and is not allocated in any category.)

It makes a big difference to the company's bottom line how much money and especially time are spent in R&D. Companies sometimes make it easy to think that time is less valuable in R&D if the product is going to have many subsequent stages to go through before it actually enters the market. With pharmaceutical products, for example, this additional time can be 5 to 10 years! In any case, the sooner the product leaves R&D, the sooner it comes out the end of the pipeline and starts generating revenues.

Can we reduce the size of the R&D component of the fixed cost while quickly providing revenue-generating products? In some companies, R&D staff generate a "damned if you do and damned if you don't" problem. Consider scientists in a start-up company, for example: you can't live without them (no product generated) and you can't live with them ("just one more experiment and I think we'll be ready"). On the other hand, once a scientist really does become a product developer—no mean feat, sometimes, but a very worthwhile goal—profitable, high-quality products come through the pipeline like clockwork. This conversion process is an important example of how to increase a company's competitiveness.

Returning to the three-dimensional plot of HealthCo's profit mountain, Figure 1 shows that, after a certain point, it becomes very expensive to make additional sales (perhaps because acquiring additional customers requires some very expensive advertising). There is, therefore, a certain region of sales and expenses where profit is greatest.

Mountaineering in R&D

If we had to choose the best place to operate, in this simplistic example, the plateau atop the mountain is that place (Figure 1). The sooner we can get there and stay there, the better. The trick, then, is to get to the top of the mountain as quickly and efficiently as possible.

(The mountain we have drawn is really much more complicated. It can be looked at from more than three directions, and there are a lot more than two factors contributing to sales and expenses at any point. For example, one contributor to sales and expenses is the analytical performance of the product. For instance, the cost and time to establish an adequate signal-to-noise ratio (S/N) adds to the expense side. On the other hand, a better S/N may increase the value of the product to the consumer.)

One reason R&D is itself an excellent model for this argument is that R&D's activities directly affect HealthCo's bottom line. Another is that R&D is already in the business of analyzing and improving performance—this is R&D's bread and butter. Much of the methodology and thinking that can be used to bolster profits throughout an organization are already in place in R&D, although they are typically focused only on science and engineering rather than on business.

Let us work backwards, then, from the goal of high-profit returns with minimal expense. To reach the profit plateau, we need to control expenses and have some sales. To have some sales we must have a market and some product to sell. We must manufacture the product. We must develop something to manufacture. We must research ideas to provide something to develop. And, of course, we must do all sorts of other things at the same time to get this product from concept to customer.

HealthCo can be viewed as the sum total of a whole collection of mountains, and we each have our mountains to climb within it. Regardless of what department of the company we may be in, and what our respective areas of expertise are, the bottom line—company profits—needs to be, for each and everyone of us, the bottom line! The way to up your bottom line is to identify the mountains and get to the top as efficiently as possible.

The Experiment Budget

If we consider R&D in terms of the number of experiments its budget allows, we can look at experiments as R&D's "products". Then there is a variable expense associated with every unit of product, that is, every experiment. We might find, for example, that, on the average, each experiment costs the company $10,000. This figure includes salaries, office space, equipment, lights, telephones, reagents, research tools, safety precautions, compliance with regulations, etc. The fewer experiments, the less the company has to spend.

Suppose that experiments also have a typical average time; then, the fewer experiments we need to do to get the necessary information to build the company's products, the less time is required. Furthermore, the sooner the product can then go on to clinical trials and the FDA, the sooner it can get approval, the sooner it can reach the market, the sooner sales can begin, the sooner revenues accrue, the sooner the break-even point is reached, and the sooner we get our bonuses and the satisfaction of seeing our products—wrought by our efforts—in the marketplace. To borrow a description from Bob Swanson, president of Genentech, suppose the experiments are leading to a product that will bring in revenues of $50 million a year, every week that we save in R&D ultimately means $1 million in revenue we would otherwise have missed.

From another perspective, a given product requires a certain minimum amount of knowledge to be acquired before the product can be created and sold. We must perform experiments and tests in order to acquire that knowledge. Ideally we should accomplish that in the shortest amount of time and with the fewest experiments, thus minimizing the experimental budget.

Accelerating R&D

One effective way to reduce the R&D time and dollar budgets is to make experiments more efficient. The following arguments, by the way, apply in any discipline, not just in biosensors.

Suppose, for an immunosensor, we are looking at the coating of a solid phase with antibody, and we are interested in getting the best signal (S) and signal-to-noise ratio (S/N) with real samples in a test device. Suppose we already have reason to believe that the signal we get depends mainly on two factors, the concentration of the antibody used in coating the solid phase ([Ab]), and the pH of the coating solution. (Of course, in reality other significant factors affect the signal and the signal-to-noise ratio.) So we do a signal-versus-[Ab] experiment at pH 7.0 and get a simple curve with a maximum: not an unusual result (Figure 3).

Now we want to see the effect of pH on the signal, so we select that [Ab] which gave us the best signal at pH 7.0, and do an experiment where we measure signal as we change the pH of the coating solution. We get an analogous result: there is a peak in the pH curve, happily, at pH 7.0 (Figure 4). These starting conditions were reasonably chosen, because we would be very happy, from a practical perspective, if they turned out, in fact, to be the best final conditions for the system. So now we are satisfied that we have found the optimum [Ab] and pH.

Would we still be satisfied if two colleagues had performed similar experiments that appeared to have pH optima at pH 8.0 and 6.0 and [Ab] optima at 0.9 and 1.1 times the best value we ourselves had obtained? What if the only thing they had done differently was to start their experiments at different initial antibody concentrations?

What if it turned out that the signals are in practice 2 times better at pH 4.50 and 2/3 the previous "best" antibody concentration? Would we have discovered that fact by using this experimental strategy? Perhaps—and would we have discovered it quickly or slowly?

Enter the mountaineers! Let's look at a two-dimensional contour map of a mountain that would show this behavior and provide more information about the true nature of the system (Figure 5). It would be easy to rationalize the described phenomena in physical terms using conventional wisdom: Denaturation of antibody can occur at very low and high pHs; inadequate coating at low [Ab] and unstable stacking of antibody molecules at high [Ab] lowers the available amount of antibody; conformational changes at moderately low pH may lead to exposure of better coating sites on the antibody, and a better angle of attachment, giving better access to antigen in the test; etc.

Our job now seems a little different. Instead of trying to find the best conditions for a single factor at a time, by sequentially sliding along individual curves, we are now trying to find the quickest route from some arbitrary starting place to the top of a mountain. The shortest route to the top of a mountain is along the line of steepest slope. (This is also the shortest route to the bottom of a mountain—the steeper the slope, the faster you get to the bottom. Also, although real-life mountain climbing must also contend with gravity, gravity is not relevant here for us experimenters: we could just as easily have chosen to start our experimental runs at the bottom or the top of the mountain, if only we had had the foreknowledge of where these were located.)

We could not tell by looking at pH or [Ab] alone where and in which direction the slope is steepest. In fact, we must look at both pH and [Ab] at once! That is, we must examine more than one factor at a time. Of course, if we look at more than one factor at a time, we won't know which one is responsible for the effect we are seeing. Nor should we, because the fact is that one factor is not independently responsible for the results we observe, as the curves show: The effect of one factor here depends on the current setting of the other factor. (Statisticians call this a factor interaction (2, 4). In this case, the physical antibody + antigen interactions measured experimentally have a fundamental correlate in the corresponding mathematical interaction terms of the equation that describes signal in terms of the antibody and antigen concentrations. Thus, the statisticians factor interaction here gives us real information about the physical world.)

Also, as product developers (rather than research scienctsts), we are not trying to perform a mechanistic study of how antibody coats solid surfaces. We are, instead,

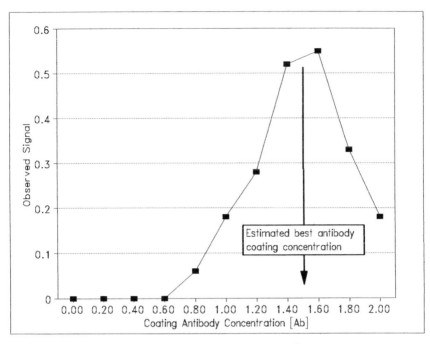

Figure 3.   Effect of [Ab] on signal, coating at pH 7.

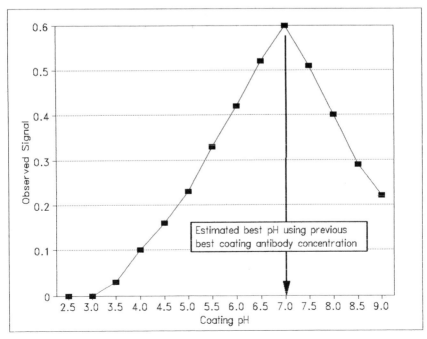

Figure 4.   Effect of coating pH on signal, using best [Ab].

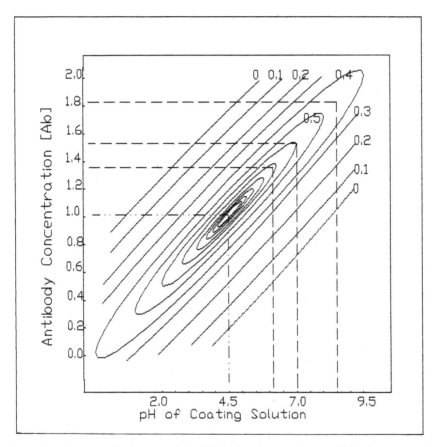

Figure 5. Contour plot from statistically designed experiment. This plot shows the effect of pH and antibody concentration on the observed signal. The contour lines are 0.1 signal units apart. The antibody concentration and signal are in arbitrary units.

interested in where the peak of the mountain is, because that's where our product is. When we get there, it may be quite suitable then to ask about mechanisms, but when we are way down the mountain slopes we have no reason to believe that the predominant mechanisms are the same there as they are at the peak. It's a question of priorities: first get to the peak; then ask how sensitive the system is to the factors; then ask what mechanisms account for that sensitivity. In many real systems, for real products, there are no hard answers, only best guesses.

In product development, the system is really much more complex. Many different factors and their interactions affect the observed signals. The one-factor-at-a-time strategy is simply inadequate. The mountaineering approach leads us to a many-factors-at-a-time strategy. Indeed, such a strategy can be used to reach the mountain top very efficiently, or to go quickly from one arbitrary location to another of our choosing.

Fortunately, many powerful methods are already established for accomplishing this strategy, such as factorial designs (*2, 4, 6*) and simplex optimization (*5*). Despite the availability of such techniques, it is astonishing that statistically designed experiments are so rarely used in analytical chemistry and product development (*7*). In practice, these techniques provide better understanding and control, faster results, lower R&D costs, and reduced time to market.

Conclusion

In summary, this discussion has examined the power of paradigms and paradigm shifts to achieve our goals. A major element of success is the search for new ways of looking at problems, identifying them, defining them, and solving them. Some tools can solve our problems even though we thought they weren't intended for us. Mountaineering is just one of many potentially valuable paradigms, applicable throughout the company, for improving competitive position. Finding new paradigms is the route to the mountain top.

References:

1. Kuhn, T. S. *The Structure of Scientific Revolutions*; 2nd ed., University of Chicago Press: Chicago, IL, 1970.
2. Natrella, M. G. *Experimental Statistics*; National Bureau of Standards Handbook 91, U.S. Government Printing Office: Washington, D.C., 1966, pp 1–17.
3. Kimball, A. W. "Errors of the Third Kind in Statistical Consulting", *J. Am. Stat. Assoc.*, **1957**, *57*, 133.
4. Deming, S.; Morgan, S. "Simplex Optimization of Reaction Yields", *Science,* **1975**, *189*, (4025), 805–806.
5. Box, G. E. P.; Hunter, W. G.; Hunter, J. S. *Statistics for Experimenters: An Introduction to Design, Data Analysis, and Model Building*; John Wiley & Sons: New York, 1978.
6. Murray, J. S., Jr. *X-Stat. Statistical Experimental Design/Data Analysis/Nonlinear Optimization*; John Wiley & Sons: New York, 1984.
7. Penzias, A. "Teaching Statistics to Engineers", *Science,* **1989**, *244* (4908), 1025.

RECEIVED August 21, 1992

Author Index

Affiliation Index

Subject Index

Production: Margaret J. Brown
Indexing: Colleen P. Stamm
Acquisition: Cheryl Shanks
Cover design: Susan Schafer

Printed and bound by Maple Press, York, PA

Highlights from ACS Books

Good Laboratory Practices: An Agrochemical Perspective
Edited by Willa Y. Garner and Maureen S. Barge
ACS Symposium Series No. 369; 168 pp; clothbound, ISBN 0–8412–1480–8

Silent Spring Revisited
Edited by Gino J. Marco, Robert M. Hollingworth, and William Durham
214 pp; clothbound, ISBN 0–8412–0980–4; paperback, ISBN 0–8412–0981–2

Insecticides of Plant Origin
Edited by J. T. Arnason, B. J. R. Philogène, and Peter Morand
ACS Symposium Series No. 387; 214 pp; clothbound, ISBN 0–8412–1569–3

Chemistry and Crime: From Sherlock Holmes to Today's Courtroom
Edited by Samuel M. Gerber
135 pp; clothbound, ISBN 0–8412–0784–4; paperback, ISBN 0–8412–0785–2

Handbook of Chemical Property Estimation Methods
By Warren J. Lyman, William F. Reehl, and David H. Rosenblatt
960 pp; clothbound, ISBN 0–8412–1761–0

The Beilstein Online Database: Implementation, Content, and Retrieval
Edited by Stephen R. Heller
ACS Symposium Series No. 436; 168 pp; clothbound, ISBN 0–8412–1862–5

Materials for Nonlinear Optics: Chemical Perspectives
Edited by Seth R. Marder, John E. Sohn, and Galen D. Stucky
ACS Symposium Series No. 455; 750 pp; clothbound; ISBN 0–8412–1939–7

Polymer Characterization:
Physical Property, Spectroscopic, and Chromatographic Methods
Edited by Clara D. Craver and Theodore Provder
Advances in Chemistry No. 227; 512 pp; clothbound, ISBN 0–8412–1651–7

From Caveman to Chemist: Circumstances and Achievements
By Hugh W. Salzberg
300 pp; clothbound, ISBN 0–8412–1786–6; paperback, ISBN 0–8412–1787–4

The Green Flame: Surviving Government Secrecy
By Andrew Dequasie
300 pp; clothbound, ISBN 0–8412–1857–9

For further information and a free catalog of ACS books, contact:
American Chemical Society
Distribution Office, Department 225
1155 16th Street, NW, Washington, DC 20036
Telephone 800–227–5558